OXFORD PHILOSOPHICAL MONOGRAPHS

Editorial Committee:
Jonathan Barnes, Michael Dummett, Anthony Kenny,
Michael Rosen, Ralph C. S. Walker

The Justification of Science and the
Rationality of Religious Belief

The
Justification of Science
and the
Rationality of Religious Belief

Michael C. Banner

CLARENDON PRESS · OXFORD

Oxford University Press, Walton Street, Oxford OX2 6DP

Oxford New York Toronto
Delhi Bombay Calcutta Madras Karachi
Petaling Jaya Singapore Hong Kong Tokyo
Nairobi Dar es Salaam Cape Town
Melbourne Auckland
and associated companies in
Berlin Ibadan

Oxford is a trade mark of Oxford University Press

Published in the United States
by Oxford University Press, New York

First published 1990
First issued in Clarendon Paperbacks 1992

British Library Cataloguing in Publication Data
Banner, Michael C.
The justification of science and the rationality of religious belief.
1. Religious beliefs. Rationality
2. Science. Theories
I. Title
200.1
ISBN 0–19–824019–8

Library of Congress Cataloging in Publication Data
Banner, Michael C.
The justification of science and the rationality of religious belief/Michael C. Banner.
p. cm.—(Oxford philosophical monographs)
Includes bibliographical references.
1. Religion—Philosophy. 2. Science—Philosophy. I. Title.
II. Series.
BL51.B238 1990   121'.7—dc20   89–22852
ISBN 0–19–824019–8

Printed in Great Britain by
Biddles Ltd, Guildford and King's Lynn

To my parents
in love and gratitude

Acknowledgements

THE title of this book records significant debts in blending the titles of two others—Basil Mitchell's *Justification of Religious Belief* and W. H. Newton-Smith's *Rationality of Science*. It was my good fortune to have Professor Mitchell's critical and conscientious guidance whilst I wrote my doctoral thesis and I acknowldege with gratitude all that I owe to him. Bill Newton-Smith was one of my tutors whilst I was an undergraduate at Balliol and was liberal then and in subsequent years in his willingness to discuss the philosophy of science with me. I hope that another tutor at Balliol, Dr P. B. Hinchliff, will not think that his kindness receives a poor return in this work.

Both Dr A. J. P. Kenny and Professor R. Swinburne generously read and commented upon sections of this work in its early drafts. Audiences at the Universities of Durham and Birmingham, at Calvin College, Grand Rapids, and at the Center for Theology and the Natural Sciences in Berkeley heard some of the ideas and were helpfully sceptical. I was also fortunate in having an opportunity to meet and debate with a number of philosophers in Czechoslovakia.

This book was all but finished whilst I enjoyed the hospitality of St Peter's College, Oxford where I held the Bampton Research Fellowship. I record my gratitude to the Electors to the Bampton Fellowship, and in particular to Professor Ernest Nicholson and the Revd W. L. R. Watson, for the encouragement given in electing me and subsequently.

To Jane Wheare, who has shared this and so much else, more than thanks are due.

M. B.

Peterhouse, Cambridge
April 1989

Contents

1.

Introduction

> Upon the whole we may conclude that the *Christian Religion*
> not only was at first attended with miracles, but even at this
> day cannot be believed by any reasonable person without one.
> Mere reason is insufficient to convince us of its veracity. And
> whoever is moved by *Faith* to assent to it, is conscious of a
> continued miracle in his own person, which subverts all the
> principles of his understanding, and gives him a determination
> to believe what is most contrary to custom and experience.
>
> (D. Hume, *An Enquiry Concerning Human Understanding*)

> Religious believers, unlike scientists, typically and character-
> istically seek to preserve their favoured models from criticism
> *at all costs* and *in the face of whatever difficulties* they
> encounter—something that would certainly be seen as ir-
> rational in a scientist.
>
> (A. O'Hear, *Experience, Explanation and Faith*)

HUME's ironic comments on religious faith express a judgement
which has become enshrined in the contrast between the rational
behaviour of the scientist and the irrational behaviour of the
religious believer, and has been commonplace since his day. It is
frequently alleged that there is a sharp dichotomy between scientific
belief and religious faith, and in particular that in science canons of
rationality are observed and honoured which are flouted or
breached by the religious believer. Whereas science presents us with
the very paradigm of sound reasoning, in religion we see a
paradigm of reasoning which is speculative or ill-founded. Indeed,
it has seemed to some that faith is tantamount to a sort of cognitive
acrobatics, whereby the mind is enabled to assent without reserve
to propositions which are in fact dubious.

The dichotomy can be established with ease if religious belief at
its worst and most superstitious is compared with scientific belief at
its best and most rigorous, but that would hardly be to the point. A
more interesting dichotomy would be demonstrated only if it could
be shown that, by the principles of rationality which typically and

properly govern the acceptance of belief in science, religious belief is irrational.

The demonstration of an interesting dichotomy is naturally no easy task. It presupposes an account of the nature and character of science as much as of the nature and character of religious belief, when in each case controversy surrounds attempted philosophical analyses. Many works in the philosophy of religion, however, seem to take for granted the character of science, as if there were some widely accepted description of the standards of scientific debate against which religious belief may be compared. But a slight acquaintance with the philosophy of science shows that there is no such agreed conception, and that whether science can be said to be rational, and if so in what sense, have been vexed questions in the field since the demise of Logical Positivism. One could almost say that apart from an all but universal welcome for the demise of Logical Positivism, there has been little consensus in the philosophy of science during the last forty years. Some, such as Kuhn and Feyerabend, have disputed the claim that science is a rational enterprise and have seemed to assert the need for a largely sociological account of scientific change. Others, such as Popper, Lakatos, Laudan, and the current Realist school, have defended the rationality of science but have differed profoundly in how they conceive of that rationality. So even were it a simple matter to give an account of the nature of religious belief, establishing a dichotomy between science and religion would still require a good deal of work in justifying the view of science which the dichotomy supposes.

Since there is no conception of science which can be taken for granted, we should be surprised at the way in which a particular view of science is sometimes relied upon, and presupposed, in the philosophy of religion. Let me take as an example one of the arguments which Don Cupitt appeals to in his attack on what he terms 'theological realism', that species of realism which supposes that the central theories of religion are in some way descriptive. Against such realism, Cupitt seeks to persuade us that 'To speak of God is to speak about the moral and spiritual goals we ought to be aiming at, and about what we ought to become. . . . The true God is not God as picturesque supernatural fact, but God as our religious ideal.'[1] Now part of his case for the rejection of traditional

---

[1] D. Cupitt, *The Sea of Faith* (London: BBC, 1984), 270.

Christian faith seems to depend on the rejection of realism as such—that is, the view that theories refer and describe. For if realism is to be rejected then religious realism is worthy of scorn just because it is realist.

In his recent *Sea of Faith*, Cupitt alleges that the realist is naïve, and that instrumentalism prevails in science. Though he devotes no noteworthy arguments to defending this assumption, it is expressed in the form of dogmatic asides which hint that the truth is known to the philosophically literate. Thus he writes:

It became increasingly clear [by the end of the nineteenth century] that all theories are not discoveries but inventions, human imaginative constructions that are imposed upon experience and can be described as 'true' only in the sense that, and for so long as it is found that, they work usefully. To put it brutally, there *is* no ready-ordered objective reality any more: there is only the flux of becoming, and the continuing ever-changing human attempt to imagine and impose order. We have to *make* sense; *we* have to turn chaos into cosmos.[2]

Essentially the same claims are repeated in *Only Human*, once again with hardly any recognition that they might be thought controversial and so require support from argument as well as from rhetoric. We are informed:

There is not any Reality or Truth in the old sense; there are only the endlessly varied visions and values that human beings project out upon the flux in order to give their lives a kind of meaning. We constructed all the world-views, we made all the theories. They are ours, and we are not accountable to them. They depend on us, not we on them. . . . No world-view, faith or ideology whatever is objectively anchored in and guaranteed by the nature of things, in the sense that human beings once believed in, for there is no such nature of things.[3]

In consequence we are assured that 'Arguments between realism and anti-realism do not after all mean much, for if the former is a non-thing, so is also the latter.'[4] The realist notion of knowledge is obsolete; all we have is our language, beyond which it makes no sense to go.

---

[2] Ibid. 188.
[3] D. Cupitt, *Only Human* (London: SCM, 1985), 9. Cupitt seems, conveniently, to forget his understanding of theories as projections to which 'we are not accountable' and his promise that 'we shall not be dogmatic about . . . scientific theories' when arguing that traditional religious doctrines are to be rejected for the reason that they conflict with scientific claims.
[4] Ibid. 51.

Cupitt is not, of course, necessarily trying to establish a dichotomy between science and religion. In his eyes the theories of both have no claim to describe the world. But what his attack on 'theological realism' illustrates is not only the variety of understandings of the nature of science in the philosophy of religion, but also the danger of treating whatever understanding is preferred as uncontroversial and as needing no support. For if one is concerned to compare the two and consider the charge that there is a significant dichotomy between science and religion, such a strategy will be unsatisfying. For then one will need not only to be sensitive to the character of religious belief, but also to take seriously the task of elucidating the nature of science, aware of the controversies which have divided philosophers of science. Only on the basis of such an elucidation can a well-founded comparison proceed.

So it is that this book has two parts, the first of which deals with questions belonging to the philosophy of science and is a necessary preliminary to the second part, the concerns of which fall within the scope of the philosophy of religion. The aim of that second part will be to arrive at an assessment of the thesis that there is necessarily a methodological dichotomy between science and religion, and that the justification of religious belief could bear no resemblance to the justification of science.

Part I begins with a consideration of the philosophy of science advanced by Thomas Kuhn. Though it cannot be regarded as satisfactory, an appreciation of its strengths and of its shortcomings provides the basis for a conception of science which is defended in Chapter 3 and elucidated in subsequent chapters. This conception of science, which I term 'rational realism', is the one against which religious belief may be compared in Part II.

PART 1

The Justification of Science

2.

Kuhn's Challenge to the Rationality
of Science

ONE of the most influential and important books in recent philosophy of science has been Thomas Kuhn's *Structure of Scientific Revolutions*. Whilst, however, there is little disagreement about its influence, its interpretation is a matter of considerable dispute. On the one hand Kuhn has made clear in his later writings that he takes himself to have been doing no more than presenting a new conception of the rationality of science. On the other hand, many have interpreted *The Structure of Scientific Revolutions* as more radical in its aim and as posing an attack on our everyday assumption that science is rational.

The question of interpretation is not the one with which I am primarily concerned. Whether Kuhn has abandoned an earlier radical position in later writings or whether he has merely resolved tensions or ambiguities in his thought in favour of a more moderate view, it is not my main purpose to determine. Rather the examination of Kuhn's work in this chapter serves as a step towards an understanding of the nature of science and scientific progress, so it is better to grasp the main lines of Kuhn's position than to agonize over the subtleties of interpretation. Whether or not Kuhn is really committed to a particular thesis is less important than the defensibility of the thesis itself.

A summary of the argument in this chapter may provide a sense of direction. What I shall be aiming to establish is that the irrationalist tendencies which have been found in Kuhn's writing are deeply unsatisfactory. The importance of appreciating the inadequacies of such an approach is in its enabling us to appreciate the particular merits of the realist philosophy of science which I shall consider and propose in Chapter 3.

It will emerge that I am inclined to interpret Kuhn's overall position as a radical one and as intended to cast doubts on the rationality of science. Even his later revisions or 'clarifications'

8 *The Justification of Science*

seem not to go far enough to entitle his position to be regarded as that of a rationalist, though they do enable Kuhn to give an explanation of a feature of science which the stress on incommensurability in *The Structure of Scientific Revolutions* renders mysterious, namely the rapid emergence of scientific consensus. But there is another noteworthy feature of science still left unexplained by Kuhn's account of the role of 'values' in directing theory choice. That is the success or progress of science, and I shall argue not only that Kuhn has no satisfactory explanation of science's success, but also that the very notion of progress in science turns out to be derivative from the concept of truth which Kuhn rejects. If an account of success is to be forthcoming, we must look to the realist philosophy of science which is the subject of Chapter 3.

1. *The Structure of Scientific Revolutions*

In order to understand *The Structure of Scientific Revolutions* one has to consider the concept of a paradigm which is central to the book's argument. Kuhn has since recognized his use of the concept to have been 'badly confused',[1] for the reader is offered such an enormous catalogue of characterizations of paradigms that the main point of introducing the concept may be lost. According to one critic, for example, the term

covers a range of factors in scientific development including or somehow involving laws and theories, models, standards, and methods (both theoretical and instrumental), vague intuitions, explicit or implicit metaphysical beliefs (or prejudices). In short, anything that allows science to accomplish anything can be part of (or somehow involved in) a paradigm.[2]

And in Kuhn's own words, a paradigm consists of a 'strong

[1] T. S. Kuhn, 'Reflections on My Critics' in *Criticism and the Growth of Knowledge*, ed. I. Lakatos and A. Musgrave (Cambridge: Cambridge University Press, 1970), 234. See also Kuhn's 'Postscript' to *The Structure of Scientific Revolutions*, 2nd edn. (Chicago: University of Chicago Press, 1970), 174–191, and 'Second Thoughts on Paradigms', in *The Essential Tension* (Chicago: University of Chicago Press, 1977), 293–319.

[2] D. Shapere, 'The Structure of Scientific Revolutions', *Philosophical Review*, 73 (1964), 385. M. Masterman claims to have found twenty-two different uses of the term; 'The Nature of a Paradigm', in *Criticism and the Growth of Knowledge*, ed. Lakatos and Musgrave, 61.

network of commitments—conceptual, theoretical, instrumental, and methodological'.[3]

In response to criticism, Kuhn has identified two aspects of scientific activity which the word 'paradigm' was supposed to capture. First of all the notion of a paradigm refers to a past scientific success which is granted an exemplary status by what Kuhn terms 'normal science'. Second, Kuhn uses the term 'paradigm' less narrowly, to describe that which constitutes the broad common ground which unites a particular group of scientists in a particular period in their practice of science. In the first sense a paradigm is an exemplar, 'the accepted way of solving a problem which serves as a model for future workers'; in the second sense, a paradigm is a 'disciplinary matrix' including, for example, the generalizations, methods, and models 'shared by those trained to carry out the work that models itself on the paradigm-as-achievement'.[4]

Now the paradigm as exemplar is but a part of the paradigm as disciplinary matrix, and it is the second and broader characterization which is more important in reading Kuhn's work. He asserts that if we are to 'understand how a scientific community functions as a producer and validator of sound knowledge' then we have to understand the operation of more than a single item in a disciplinary matrix.[5] His theory of scientific change, therefore, draws attention to paradigms in the larger, not the smaller sense.

During a period of normal science—which is to say during the period of the acceptance of a particular paradigm—prima-facie anomalies are treated not as refuting it, but as posing problems for which solutions will be forthcoming.[6] Kuhn's important insight here is that a refutation of a theory does not consist solely in a phenomenon which it cannot handle. He writes that 'if an anomaly is to evoke a crisis, it must usually be more than just an anomaly.'[7] When it comes to specifying what this something more must be Kuhn is vague, but suggests that a paradigm is threatened when

---

[3] *Structure of Scientific Revolutions*, 42.

[4] I. Hacking, editor's introduction to *Scientific Revolutions* (Oxford: Oxford University Press, 1981), 2–3. See also Kuhn, 'Second Thoughts on Paradigms', 297–8, and 'Reflections on my Critics', 271–272.

[5] 'Second Thoughts on Paradigms', 298.

[6] *Structure of Scientific Revolutions*, 77 f.

[7] Ibid. 82.

anomalies proliferate, or when one anomaly holds a particular fascination, and hence its stubbornness achieves a heightened significance. An anomaly takes on this role if it seems to call in question fundamental aspects of the paradigm, when its solution has considerable practical importance, or perhaps merely when the anomaly is long outstanding.[8] When it happens that an anomaly takes on this added significance, a crisis occurs. Crises close in different ways, but sometimes there is a transition from normal to extraordinary science, the latter characterized by the clash of paradigms. If the new contender prevails, then the crisis is the prelude to a scientific revolution.[9]

In describing the transition which the anomaly induces, why does Kuhn use the word 'revolution'? In asking that question we are led to the heart of the matter, for Kuhn employs this description to announce his opposition to what he thinks of as the standard, cumulative view of scientific change, according to which scientific theories are replaced when they are confronted by facts which render them no longer tenable. According to this standard view, Newton's theory gave way to Einstein's when the facts piled up in favour of the latter, and the theory of fixity of species gave way under a weight of evidence which Darwin assembled from his study of the Galapagos Islands and which favoured a theory of evolution. Thus common sense holds that theories follow one another in a steady progression, as the rules of science show that a new theory is better than the last.

Putting the standard view in terms of the imagery which Kuhn employs, we might say that scientific change is analogous to political change in a constitutional system. Just as the succession of monarchs, governments, and parliaments is in accord with the rules of the constitution, so too in science, theory change is ordered and disciplined, settled by well-established and accepted rules.

Against a vision of the sciences as peaceably governed by some ruling methodology, Kuhn sets a conception of theory change as revolutionary. Political revolutions occur when a change in political

---

[8] *Structure of Scientific Revolutions*, 82.

[9] Ibid. 84 f. Kuhn's assumption that the sense of crisis prior to a 'revolution' is ubiquitous, is properly challenged by M. Heidelberger in 'Some Intertheoretic Relations Between Ptolemean and Copernican Astronomy', in *Paradigms and Revolutions*, ed. G. Gutting (Notre Dame, Ind.: University of Notre Dame Press, 1980), 271–283.

power is dictated by force rather than sanctioned by the legitimating conventions of the previous authority.[10] In other words, it is in the nature of a political revolution that it is 'extra-constitutional', for the succession of authorities is not determined in accordance with accepted standards, but itself establishes new ones. Now it is Kuhn's contention that 'the historical study of paradigm change reveals very similar characteristics in the evolution of the sciences.'[11] In particular, he claims that when scientific revolutions occur, the authority of the victorious theory is not recognized by the loser, for it is not 'out-argued', but overthrown. The point is made in the following words:

Like the choice between competing political institutions, that between competing paradigms proves to be a choice between incompatible modes of community life. Because it has that character, the choice is not and cannot be determined merely by the evaluative procedures characteristic of normal science, for these depend in part upon a particular paradigm, and that paradigm is at issue. When paradigms enter, as they must, into a debate about paradigm choice, their role is necessarily circular. Each group uses its own paradigm to argue in that paradigm's defense.[12]

Here it seems that the protagonists in a scientific dispute are, like those in revolutionary political conflict, talking through one another. However, Kuhn does not deny that argument in such a context may be compelling, but alleges rather that it is compelling only in a way which he dubs 'persuasive'. Wittgenstein once remarked: 'At the end of reasons comes *persuasion*. (Think what happens when missionaries convert natives.)'[13] Kuhn similarly gives 'persuasive' a pejorative tone, meaning to draw attention to the fact that the argument

cannot be made logically or even probabilistically compelling for those who refuse to step into the circle. The premises and values shared by the two parties to a debate over paradigms are not sufficiently extensive for that. As in political revolutions, so in paradigm choice—there is no higher standard than the assent of the relevant community.[14]

---

[10] *Structure of Scientific Revolutions*, 93 f.
[11] Ibid. 94.
[12] Ibid.
[13] L. Wittgenstein, *On Certainty*, ed. G. E. M. Anscombe and G. H. von Wright, trans. D. Paul and G. E. M. Anscombe (Oxford: Blackwell, 1969), para. 612.
[14] *Structure of Scientific Revolutions*, 94.

It is this last comment, that 'there is no higher standard than the assent of the relevant community', which challenges most strongly the traditional conception. Surely the evaluative procedures of science are what determine the winner in a competition between theories? The 'relevant community' holds the last word only if there are no 'higher standards', for if there is such a standard, talk of 'stepping into the circle'—a phrase reminiscent of Tillich and Barth—would be inappropriate. According to the common-sense view, of course, the assent of the community is dictated by certain agreed standards, enabling us to say that the preferred theory is the better one. But Kuhn turns this upside down. It is not a higher standard which determines the community's assent, but the community's assent which dictates what is to count as the highest standard. For Kuhn the paradigm is, so to say, the basic unit of commitment.

Whilst Kuhn's conception of scientific change is, I hope, clear, it remains to explain why he advocates it, why he talks of scientific revolutions where others have been content to talk of scientific progress. In his own account of how he came to write *The Structure of Scientific Revolutions*, Kuhn gives a central place to his study of the history of science. But in fact the crucial arguments are found to be philosophical, not historical, and it is these philosophical arguments which are the key to the book. And amongst these arguments it is those which relate to his defence of the incommensurability thesis which are most important. For, Kuhn tells us:

Just because it is a transition between incommensurables, the transition between competing paradigms cannot be made a step at a time, forced by logic and neutral experience. Like the gestalt switch, it must occur all at once (though not necessarily in an instant) or not at all. . . . in these matters neither proof nor error is at issue. The transfer of allegiance from paradigm to paradigm is a conversion experience that cannot be forced.[15]

Kuhn's understanding of science as governed by the mores of revolutionary politics derives from the thesis that paradigms are in some sense incommensurable. But what exactly does Kuhn mean by 'incommensurable'? Literally, 'incommensurable' means lacking a common measure and so a natural construal of Kuhn's claims is that scientific theories, like competitors for political power, lack a

---

[15] *Structure of Scientific Revolutions*, 150–1.

common measure by which we might decide between them. Because there is no such measure, the comparison of one with another is impossible, and hence it is impossible for one to have a rational preference. And if rational preference is out of the question, then the standard view of science as rule-governed and constitutional cannot hold. Believing in the incommensurability of paradigms, Kuhn is necessarily sent in search of external reasons for scientific change. But this explanation only pushes the question one step further back. If Kuhn thinks that science is revolutionary because he accepts the doctrine of incommensurability, we are led to ask why he advocates that doctrine.

Several lines of thought in *The Structure of Scientific Revolutions* seem to provide a basis for Kuhn's position:

(1) that there will be inevitable disagreement between the proponents of different paradigms about the goals or standards of science: 'In learning a paradigm the scientist acquires theory, methods, and standards together, usually in an inextricable mixture. Therefore, when paradigms change, there are usually significant shifts in the criteria determining the legitimacy both of problems and of proposed solutions.'[16]

(2) that communication across the revolutionary divide is at best partial because of changes of meaning in central theoretical terms.[17]

(3) that different paradigms produce different worlds: 'In a sense that I am unable to explicate further, the proponents of competing paradigms practice their trades in different worlds.'[18]

Of these three sources, the second seems to be the most important. Kuhn has insisted in his later work that he should not be read as asserting (1); he is not claiming that paradigms determine the standards used for choosing between theories.[19] Change of paradigm occurs for good reason, a principle which is said to be

---

[16] Ibid. 109; see also pp. 103 and 148. In his account of Kuhn's philosophy of science, Laudan notes Kuhn's emphasis on the holistic nature of scientific change and his taking for granted that ontological differences between theories will be accompanied by methodological and axiological divergence. See L. Laudan, *Science and Values* (Berkeley: University of California Press, 1984).

[17] *Structure of Scientific Revolutions*, 101–2 and 128.

[18] Ibid. 150.

[19] 'Reflections on my Critics', 261.

found in chapter 12 of *The Structure of Scientific Revolutions*, where there is a preliminary formulation of those good reasons. Unfortunately the chapter to which Kuhn refers is by no means unequivocal on this point. Though we are told that scientists do not transfer from one paradigm to another on the ground of 'some mystical aesthetic', we are informed that the decision to switch from old to new is made 'on faith', on a basis which 'need be neither rational nor ultimately correct'.[20] Far from defending the existence of good reasons for theory choice, the chapter under discussion seems to suppose that there are no such shared standards.

Whether or not Kuhn should be read as denying the constancy of standards by which theories are judged, there are two other lines of thought which independently of this would, as I have mentioned, support Kuhn's claim that in the debate over paradigms 'there is no higher standard than the assent of the relevant community.' These are (3) that paradigms are 'constitutive of nature',[21] and (2) that communication between paradigms is partial. For both of these there is undoubted textual support and each would provide sufficient explanation for Kuhn's position on incommensurability and his analogy between scientific change and political upheaval.

Scientists, we are told, see the world through a paradigm with the consequence that 'the proponents of competing paradigms practice their trades in different worlds.' As an example of this phenomenon Kuhn notes that after Dalton 'chemists came to live in a world where reactions behaved quite differently from the way they had before.'[22] This comment may be read in two ways. In one sense it seems to espouse the idealist doctrine that the world is a construction of our minds and theories, a doctrine from which incommensurability follows without more ado. For if each theory creates its own world, there is certainly no common ground by reference to which theories might be compared.

The same comments may, however, be given another sense, ridding them of idealist assumptions and rendering them unobjectionable. Thus we might say, in a certain manner of speaking, that after Lavoisier, for example, scientists lived in a different world,

[20] *Structure of Scientific Revolutions*, 158.
[21] Ibid. 110.
[22] Ibid. 134.

one with oxygen rather than with dephlogisticated air. This is to say no more than we would in saying of Europeans who lived prior to Columbus's voyages, that they lived in a world which did not include America.[23] In this sense we would agree with Kuhn that after a scientific revolution 'the whole network of fact and theory . . . has shifted.'[24]

Should we read Kuhn as asserting the idealist thesis that the world is a creation of our theories, or more mildly, as asserting that the world is a different place for those who have discovered more about it? That Kuhn wishes to be read in a mild, non-idealistic way is the burden of later disclaimers,[25] which point to the fact that much of what he writes in *The Structure of Scientific Revolutions* is incompatible with idealism in its full-bodied form. Thus Kuhn comments that 'it is hard to make nature fit a paradigm. That is why the puzzles of normal science are so challenging.' And elsewhere we are told: 'Normal science does and must continually struggle to bring theory and fact into closer agreement.'[26]

Having set on one side notions about variations in standards, and the more outlandish idealistic reading of Kuhn's 'different worlds' terminology, we still lack anything which would justify the radical claim which we are seeking to understand, namely: that the assent of the scientific community is the highest standard in disputes over paradigms. It remains, however, to consider the thesis concerning the limitations on communication across paradigms, which seems obliged to bear the weight of Kuhn's theory.

Kuhn's sense of the limits on translation derives from a theory of meaning, for given this theory the different worlds of which Kuhn speaks can be understood as different worlds of meaning.[27] According to Newton-Smith the theory in question is a relic of a positivist approach.[28] Essentially it holds that the meaning of an

---

[23] H. Meynell, 'Truth, Witchcraft and Professor Winch', *Heythrop Journal*, 13 (1972), 162.

[24] *Structure of Scientific Revolutions*, 141.

[25] For a denial of idealism see 'Reflections on my Critics', 263.

[26] *Structure of Scientific Revolutions*, 135 and 80.

[27] For the later Kuhn a similar doctrine is rooted in a neo-Quinean belief in the indeterminacy of translation; see *The Essential Tension*, pp. xxii–xxiii.

[28] W. H. Newton-Smith, *The Rationality of Science* (London: Routledge & Kegan Paul, 1981), 9–13 and ch. 7. See also C. R. Kordig, *The Justification of Scientific Change* (Dordrecht: Reidel, 1971), ch. 2.

individual term in a theory is given by its relationship to the
structure of the entire theory. Since Kuhn denies an observation/
theory dichotomy,[29] this principle applies to all terms employed in
scientific discourse. It follows that what the Newtonian says about
mass and what the Einsteinian says about mass are not said about
the same thing, and further, that one cannot generate a clash
between them at the observational level since those terms which the
positivists construed as observational are now understood to be
theory-laden and hence to change their meaning with change in
theory. The Copernican who refused to call the sun a planet not
only saw a new world, but also gave the word 'planet' a new
meaning,[30] just as the sentence p asserted by Einstein must differ in
meaning from the sentence p denied by Newton. Since the theories
are different, no word can translate from one to the other and hence
there can be no disagreement between them which is not verbal
equivocation. Nor, if the two scientists agree in asserting q, are they
to be taken to be in real agreement. Thus on the basis of this theory
of meaning, Kuhn denies the usual derivation of Newtonian from
Einsteinian dynamics, subject to limiting conditions.[31]

The 'radical meaning variance', as Kordig dubs it, which is at the
heart of Kuhn's understanding of science, drives us to the
conclusion that paradigms are incommensurable. Since they talk
about different things and cannot be translated, there is no
possibility of comparing them. (How they manage at the same time
to be incompatible—as Kuhn holds them to be—remains a
mystery.) And so Kuhn is led to liken a change in allegiance
between paradigms to a 'conversion' and to say that 'the switch of
gestalt . . . is a useful elementary prototype for what occurs in full
scale paradigm shift.'[32]

Just because it is a transition between incommensurables, the transition
between competing paradigms cannot be made a step at a time, forced by
logic and neutral experience. Like the gestalt switch it must occur all at
once (though not necessarily in an instant) or not at all.[33]

[29] *Structure of Scientific Revolutions*, 125 f.
[30] Ibid. 128–9.
[31] Ibid. 101–2; and see I. Hacking, *Representing and Intervening* (Cambridge:
Cambridge University Press, 1983), 72–4.
[32] *Structure of Scientific Revolutions*, 85.
[33] Ibid. 150.

If there is no possibility of rational comparison of paradigms, and proof and error are not at stake, what occurs is simply a change, as Kuhn acknowledges.[34] And since there is no higher standard than the assent of the relevant community we may 'have to relinquish the notion, explicit or implicit, that changes of paradigm carry scientists and those who learn from them closer and closer to the truth.'[35] So it is that *The Structure of Scientific Revolutions* closes with the moral 'that scientific progress is not quite what we have taken it to be', with Kuhn offering as a type for scientific development the image of non-teleological evolution, that is, evolution the character of which cannot be explained by some underlying goal.[36]

We asked originally what Kuhn takes to justify the political analogy which expresses his main thesis. The answer was a belief in incommensurability, and we have now examined the three lines of thought in *The Structure of Scientific Revolutions* which seek to account for the existence of incommensurability. These were the suggestions that there is a change of standards with change of paradigm, that communication across paradigms is of a limited nature, and that advocates of different theories live in different worlds. I noted that the third belief is not expounded unambiguously in Kuhn's text, and that the first is one which Kuhn goes out of his way to repudiate in his later writings. The main support, therefore, for the revolutionary conception of science derives from Kuhn's belief in that incommensurability which exists between theories whose advocates cannot fully communicate. Given the theory of meaning which Kuhn holds, it is irrelevant whether or not the standards of choice stay the same. For it would be impossible to compare theories which talk about different things whether or not those who espouse them could agree on which principles to employ in that comparison.

To conclude, the picture of science which is found in Kuhn's earlier writings supports the radical contention that there can be no comparison of paradigms, and hence that change of paradigm is beyond rational adjudication. Science, according to Kuhn, does not exemplify the rational and steady progress which is often attributed to it.

[34] Ibid. 151.　　　　[35] Ibid. 170.　　　　[36] Ibid. 170–3.

2. *Kuhn's Retreat*

It has seemed to his critics that Kuhn does not really argue for the theory of meaning on which the crucial thesis of incommensurability depends, but rather assumes it. Thus Shapere objects:

Kuhn's argument amounts simply to an assertion that despite the derivability of expressions which are in every formal respect identical with Newton's Laws, there remain differences of 'meaning'. What saves this from begging the question at issue? . . . one might equally well be tempted to say that the 'concept' of mass (the 'meaning' of 'mass') has remained the same (thus accounting for the deducibility) even though the *application* has changed.[37]

Since Kuhn neither expounds nor defends the theory of meaning which underlies his claims, Shapere's charge is largely just.

It might be thought, therefore, that the proper strategy for one suspicious of Kuhn's radical position would be to attempt a refutation of the theory of meaning upon which the position depends. But though such a refutation must be part of a detailed criticism of Kuhn's work, it is better at this stage to draw attention to the muddle into which Kuhn is led by the theory, and his consequent tacit abandonment of it. (In the next chapter I shall consider an account of the meaning of theoretical terms which the realist might develop.) For it becomes clear that though Kuhn's *Structure of Scientific Revolutions* can be read, as I have suggested, as a radical critique of what he terms the standard view of science, his most important papers since that work have been concerned to deny that interpretation. In other words, Kuhn rejects that interpretation which justifies the charge that he made science out to be irrational.[38] What most concerns us, however, is not so much the adjudication of the exegetical question, but more a review of Kuhn's reasons for dissatisfaction with the radical view (whether or

[37] '*Structure of Scientific Revolutions*', 390.
[38] The papers are Postscript to *Structure of Scientific Revolutions*, 'Logic of Discovery or Psychology of Research?' and 'Reflections on my Critics', in *Criticism and the Growth of Knowledge*, ed. Lakatos and Musgrave, and 'Objectivity, Value Judgment, and Theory Choice' and 'Second Thoughts on Paradigms' in *The Essential Tension*.

not he once advocated it) and a consideration of the picture of the scientific enterprise to which he now points us.

Imre Lakatos accuses Kuhn of having made science 'a matter for mob psychology'.[39] Putting the same point in more measured tones, one can say that a significant problem with Kuhn's *Structure of Scientific Revolutions* is its seeming inability to cope with the fact of the formation of consensus within science.

Laudan draws attention to this in his *Science and Values*, arguing that though Kuhn provides us with a sufficient explanation of the disagreements which occur amongst scientists, we also need some explanation of the ability of the scientific community to resolve its disagreements so speedily. Whether or not we describe the change from Newtonian to Einsteinian physics as a paradigm shift, we have to allow that a consensus in favour of the latter emerged rapidly. In fact, the speed with which basic ideas change in science is one of its remarkable characteristics, and cannot be explained by the various extra-scientific tactics (control of journals, appointments, and so on) which, according to Kuhn, are employed by the scientific community. Such an explanation is inadequate for the simple reason which Laudan gives, that it fails to account for how the opposition itself manages to unite around a new paradigm.[40]

Kuhn's alleged clarifications of *The Structure of Scientific Revolutions* deal with this phenomenon of rapidly emerging consensus, for they portray it as a book not concerned to question the very possibility of scientific rationality, but devoted to exploring its nature. The point, he claims, was to show that in theory choice recourse is not to proof, as it might be in mathematics or logic, but to persuasion.[41]

Here 'persuasion' has lost its pejorative tone, for the conversion that persuasion is intended to induce is subject to the possession of good reasons, a preliminary codification of which is said to be found in *The Structure of Scientific Revolutions*. These good reasons, we are told, are 'of exactly the kind standard in philosophy

---

[39] I. Lakatos, 'Falsification and the Methodology of Scientific Research Programmes', in *Criticism and the Growth of Knowledge*, ed. Lakatos and Musgrave, 178.

[40] Laudan, *Science and Values*, 16–22.

[41] 'Reflections on my Critics', 260–1.

of science: accuracy, scope, simplicity, fruitfulness, and the like'.[42] Kuhn sums up his point so:

> What I am denying then is neither the existence of good reasons nor that these reasons are of the sort usually described. I am, however, insisting that such reasons constitute values to be used in making a choice rather than rules of choice. Scientists who share them may nevertheless make different choices in the same concrete situation.[43]

What makes it appropriate, though, to describe scientists as guided by values rather than bound by rules? Two factors play a part, Kuhn suggests: collectively the criteria of theory choice may be in conflict, and individually they may be imprecise.

First, different values in the range which make up the criteria of theory choice may 'dictate . . . different choices'. For example, of two competing theories one might be simpler, the other more accurate. Different practitioners will choose different theories in accord with how they value simplicity as against accuracy. This may result in different decisions, each of which can be regarded as legitimate.

The second factor Kuhn thinks more important. Though scientists may share principles of choice, there is room for them to disagree in their application of them. Take simplicity again. What one scientist judges to be simple may not be so according to another understanding of the same criterion.[44] Or take the requirement that a theory be fruitful in explaining the phenomena within the field. Kuhn points out that the oxygen theory was acknowledged to account for the observed weight relations in chemical reactions, unlike the phlogiston theory, but that the phlogiston theory could account for metals being much more alike than the ores from which they were composed.[45] Which is the more fruitful? Scientists may 'differ in their conclusions without violating any accepted rule',[46] just as judges might differ in determining whether a certain degree of force, noise, or whatever, is reasonable.

In spite of this possibility of divergence in the application of a criterion of choice and in the ranking of the various criteria, Kuhn does not rule out that canons of choice may be discoverable. What

---

[42] 'Reflections on my Critics', 261; see also 'Objectivity, Value Judgment, and Theory Choice', 321–2.

[43] 'Reflections on my Critics', 262.

[44] Ibid.; see also 'Objectivity, Value Judgment, and Theory Choice', 322–5.

[45] 'Objectivity, Value Judgment, and Theory Choice', 323.

[46] 'Reflections on my Critics', 262.

has to be remembered, though, is that such canons 'are not by themselves sufficient to determine the decisions of individual scientists. For that purpose the shared canons must be fleshed out in ways that differ from one individual to another.'[47] But though Kuhn opposes any expectation of finding some sort of algorithm which will determine theory choice in every conceivable case, he does not deserve the criticism that he holds theory choice to be irrational. He has not made science 'subjective' if that is understood to mean 'a matter of taste'; his concern has been to highlight science as 'judgemental'.[48] Thus, as I said, Kuhn's claim is that *The Structure of Scientific Revolutions* has to do with the nature of scientific rationality, not with its very possibility.

The view to which Kuhn now holds is not one from which those who take science to be rational have cause to dissent. His claim is that 'In a debate over choice of theory, neither party has access to an argument which resembles a proof in logic or formal mathematics',[49] and that in consequence there will, in some cases, be an element of discretion in the scientist's judgement. Scientists may give priority to different criteria amongst those which govern theory choice, or they may, in applying an agreed principle, reach different conclusions because they interpret it in different ways. But this would only threaten the rational character of science if we assumed that rationality consists in following the dictates of rules which have no place for judgement or discretion. But this would be an improper assumption. The very application of rules itself requires the exercise of judgement or discretion in difficult cases, unless one is to posit an infinite regress of rules which themselves determine the application of the rules. Thus the exercise of judgement in the choice between scientific theories does not call in question the rationality of science, for it can be seen that such judgement is integral to rationality.[50]

Kuhn's considered conception of the selection of a theory as sometimes underdetermined by the principles of choice poses no threat to the rational view of science. As if to underline this point Kuhn takes up again the evolution image with which *The Structure*

[47] 'Objectivity, Value Judgment, and Theory Choice', 325; see also 'Reflections on my Critics', 241.
[48] 'Objectivity, Value Judgment, and Theory Choice', 337.
[49] 'Reflections on my Critics', 260.
[50] This point is made by B. G. Mitchell in his discussion of Kuhn's work in *The Justification of Religious Belief* (London: Macmillan, 1973), 89 f.

of Scientific Revolutions ends, asking us to picture an 'evolutionary tree representing the development of scientific specialities from their common origin'. On this tree a line runs, not doubling back on itself, from the base of the trunk to the tip of the branch, and theories found on this line are related by descent. Kuhn continues:

Now consider two such theories, each chosen from a point not too near its origin. I believe it would be easy to design a set of criteria—including maximum accuracy of predictions, degree of specialization, number (but not scope) of concrete problem solutions—which would enable any observer involved with neither theory to tell which was the older, which the descendant. For me, therefore, scientific development is, like biological development, unidirectional and irreversible. One scientific theory is not as good as another for doing what scientists normally do. In that sense I am not a relativist.[51]

The importance of this analogy is that here Kuhn seems to imply that the history of science is a history of development, not merely of change. Given this understanding of the nature of theory choice, the former talk of gestalt switches and so on becomes rather inappropriate; so too is the title, *The Structure of Scientific Revolutions*. If we accept Kuhn's elucidations as elucidations, and not as revisions, then the book would have been better named *The Structure of Scientific Judgement*, for that would have more accurately reflected what he now claims to be its central point, that is, the non-algorithmic nature of the choice between theories.

It is clear, then, that with these elucidations or modifications Kuhn has done enough to account for that first aspect of science which we noted, namely the formation of consensus. If scientific theories could not be compared, the fact that scientists ever come to agree to support one in preference to a rival would be truly remarkable. But if one sees science as governed by values or criteria of theory choice—even if vague ones—the formation of consensus is explicable. It occurs because there is a set of values which scientists apply in taking their decisions. The existence of these values, which they apply in common, explains why they come to common conclusions in most cases.

What, then, of incommensurability? In recent papers Kuhn seems to have softened his account of its significance, and has implicitly repudiated the theory of meaning on which his radical conclusions

[51] 'Reflections on my Critics', 264.

depended. True enough, he still holds incommensurability to be a noteworthy feature of scientific change. Thus in 'Reflections on my Critics', for example, Kuhn writes:

In the transition from one theory to the next words change their meanings or conditions of applicability in subtle ways. Though most of the same signs are used before and after a revolution—e.g. force, mass, element, compound, cell—the ways in which some of them attach to nature has somehow changed. Successive theories are thus, we say, incommensurable.[52]

But the notion of incommensurability is now taken up and explicated in terms of the difficulties associated with translation.[53] And the important point is that difficulties with translation are rarely insurmountable, and in any case are difficulties for specific terms within a language, rather than for the language as a whole.

The weakening of incommensurability is most likely the result of Kuhn's confronting the following dilemma: either change of theory can be explicated in terms of reasons (albeit non-determinative reasons or values) or it cannot. If it cannot, then there is no convincing explanation of the formation of consensus. If it can, then to speak of incommensurability and of conversion is no more than a dramatic way of stating the fairly modest point that scientists live in a different world after a change of paradigm, just as Europeans lived in a different world after the discovery of America. Of course, we experience psychological difficulties in trying to step from one framework to another. Accustomed to looking at the world in one way, we may for some time be blind to the merits of an alternative perspective. But what is significant is that this sort of incommensurability renders comparison difficult, not impossible.[54] In the earlier writings the incommensurability was more drastic. It prevented the comparison of theories one with another, and so was the basis for the rejection of a rational account of scientific change. The modification in Kuhn's understanding of incommensurability at least allows the possibility of a more acceptable conception of science which can take account of the formation of consensus.

That Kuhn now portrays science as *governed* by values explains

---

[52] Ibid. 266–7.

[53] Translation is also given as the type of the problems surrounding theory choice in 'Objectivity, Value Judgment, and Theory Choice', 338–9.

[54] For evidence that this is the type of 'incommensurability' which Kuhn now believes in see 'Reflections on my Critics', 232 and 276.

why I did not trouble to refute the doctrine of incommensurability
by trying to find a refutation of the theory of meaning on which it
depends. For though incommensurability was isolated as the key
element in Kuhn's commitment to the irrationality of science, his
subsequent recognition of principles of choice in the selection of
scientific theories allows one to assume that he has, quite rightly,
repudiated incommensurability. For incommensurability is incompatible with the regular emergence of consensus within science,
and any theory of meaning which implies that theories are
incommensurable is refuted by this fact alone. Should we conclude,
then, that with this qualification Kuhn has given an adequate
account of science? It is to this question that I now turn.

3. Kuhnian Science: Rational or Irrational?

I have suggested that Kuhn's attitude to incommensurability should
be construed by reference to the logic of his general position, rather
than from his asides. That general position suggests the rejection of
incommensurability and might be thought to show that Kuhn holds
science to be rational. It is important, however, to see that the
rationality Kuhn attributes to science is of a very limited sort, so
limited in fact that, even noting the rejection of incommensurability, his philosophy of science still amounts to a radical attack on
a rationalist position.

   Our use of the word 'rational' involves a certain ambiguity,
highlighted by the fact that we might describe the behaviour of a
witch doctor or of an astrologer as rational, even though we think
that the theories which their actions presuppose are deeply
mistaken. Thus, to take an example, for the medieval astrologer, in
whose world view comets were omens of divine ill will, it would be
rational to delay a journey because of a comet's appearance in the
heavens. And we would properly describe that as a rational
decision, and contrast it with the behaviour of another astrologer
who, holding the same world view, none the less set out on a
journey.

   It is the possibility of decisions being rational or irrational in this
way which allows us to explain the formation of consensus within a
discipline without needing to appeal to the use of force or pressure.
A consensus amongst astrologers is explicable just because there are

reasons and rules to which reference is made in settling questions and disputes.

There is, however, another way in which we use the words 'rational' and 'rationality'. Immunization of children, for example, we regard as a rational policy, not merely because it is a practice governed by certain specifiable rules, values, and procedures, but because it is based on a theory of the workings of the human body which we take to be true. Thus the rationality of immunization explains not only the consensus which would exist between its practitioners on some question, but would also explain the success of the practice of immunization. The important point is that immunization, as against astrology or witchcraft, possesses a rationality which accounts not only for agreements, but also for the success of an agreed policy.

The rationality which Kuhn allows to science in rejecting incommensurability is the rationality of rule-governed behaviour. Decisions in science are not based on the whims of scientists, but are guided by a conception of the qualities of a good theory. But what of the rationality of the second type, the rationality possessed by an action or choice which is based upon a theory which is true or approximately true? It is in attributing this sort of rationality to science that the realist position to be discussed in the next chapter is to be distinguished from Kuhn's.

The realist is led to describe science as, let us say, strongly rational, by attention to its success. For the fact of this success is striking. It has enabled us to anticipate and manipulate the world in various and astonishing ways, from landing on the moon to practising *in vitro* fertilization. According to the realist, if we are to explain this remarkable and increasing success, then we must attribute to science the rationality of being based on theories of increasing verisimilitude.

Kuhn is not blind to the success of science, nor unaware of the need to explain it. He writes towards the end of *The Structure of Scientific Revolutions*: 'There remains the problem of understanding why progress should be so noteworthy a characteristic of an enterprise conducted with the techniques and goals this essay has described.'[55] Eschewing the realist's answer, Kuhn is tempted to think that there could be a sociological account of scientific

[55] *Structure of Scientific Revolutions*, 162.

success. If this 'explanation' is rejected, as I think it must be, we find it hinted that the success of science consists in problem-solving, and that this success may be explained as the result of the articulation of problem-solving techniques. I shall argue, however, that this second explanation relies on, rather than avoids, the realist position. But for the moment we might consider why Kuhn wants to avoid the path followed by the realist.

Kuhn has doubts about the realist framework, so that whilst the realist will assert, with at least an intuitive plausibility, that scientific method has generated a succession of theories increasing in verisimilitude, Kuhn will have none of it. His verdict on realism is given in his paper 'Logic of Discovery or Psychology of Research?', where he asks what it is about science which requires explanation and answers that it is: 'not that scientists discover the truth about nature, nor that they approach ever closer to the truth. Unless, as one of my critics suggests, we simply define the approach to truth as the result of what scientists do, we cannot recognize progress towards that goal.'[56]

Elsewhere he writes: 'There is, I think, no theory-independent way to reconstruct phrases like "really there"; the notion of a match between the ontology of a theory and its "real" counterpart in nature now seems to me illusive in principle.'[57] While Kuhn will agree that a later theory is a better tool for the practice of normal science, he refuses to go forwards with others who,

Granting that neither theory of a historical pair is true . . . none the less seek a sense in which the later is a better approximation to the truth. I believe that nothing of that sort can be found. On the other hand, I no longer feel that anything is lost, least of all the ability to explain scientific progress, by taking this position.[58]

His conclusion is that ' "truth" may, like "proof", be a term with only intra-theoretic applications.'[59] Not only can we not recognize the truth, but the very notion is 'illusive in principle'. Hence we cannot talk of one theory as more approximately true than another, whether we are comparing theories within science or, let us say, the theories of science with the beliefs of the astrologer.

[56] 'Logic of Discovery or Psychology of Research?', 20.
[57] Postscript, 206.
[58] 'Reflections on my Critics', 265.
[59] Ibid. 266.

Doubts about truth provide the basis for Kuhn's rejection of the realist's explanation of the success of science and the realist's understanding of its rationality. Against the realist he asserts that the explanation of scientific progress 'must, in the final analysis, be psychological or sociological. It must, that is, be a description of a value system, an ideology, together with an analysis of the institutions through which that system is transmitted and enforced.'[60] Elsewhere he tells us that his 'position is intrinsically sociological'.[61]

In part this sounds right. No explanation of scientific success can ignore the 'value system' of science, if that means its methodology, for one cannot explain success by attributing to science a method which it does not possess. Scientific success can only be explained by an analysis of how it does in fact operate. But Kuhn seems to mean more than just that. One critic, echoing the Euthyphro dilemma, has asked whether the criteria of theory choice accepted by the scientific community are rational because they are the values accepted by the scientific communities, or whether they are accepted because they are the criteria of rationality.[62] For Kuhn the answer seems to be that the criteria are rational because they are accepted by the scientific community. As Wittgenstein might put it, we may say only that this game is played.

But if we follow the sociology of science which has been inspired by Kuhn's suggestions and construe rationality as consisting in the observance of those rules which the scientific community respects, we find ourselves unable to explain scientific progress.[63] Of course, the realist is not opposed to sociology as such and allows that there can be an interesting sociology of knowledge in general and of science in particular. For example, the sociologist may have interesting things to say about the provenance of certain ideas, how the development of theories is affected by their perceived economic value, their relationship to religious ideas, and so on. But the realist

---

[60] 'Logic of Discovery or Psychology of Research?', 21.
[61] 'Reflections on my Critics', 238.
[62] R.J. Bernstein, *Beyond Objectivism and Relativism* (Oxford: Blackwell, 1983), 58.
[63] Kuhn's influence is particularly evident in the work of the Edinburgh School. See B. Barnes and D. Bloor, 'Relativism, Rationalism and the Sociology of Knowledge', in *Rationality and Relativism*, ed. M. Hollis and S. Lukes (Oxford: Blackwell, 1982), 21–47; also D. Bloor, *Wittgenstein: A Social Theory of Knowledge* (London: Macmillan, 1983).

does dispute the contention that sociology explains the success or progress of science merely by describing the various interests which have determined theory choice. For part of science's progress consists in the increasing manipulative power of modern science, leaving us to wonder why the growth of this power has been consequent upon the particular, ideologically motivated predilections of scientists. Far from explaining scientific progress, sociology leaves it a mystery. As Boyd has said, 'The instrumental reliability of particular scientific theories cannot be an artifact of the social construction of reality.'[64]

Recall Kuhn's confidence that in rejecting the realist framework nothing 'is lost, least of all the ability to explain scientific progress'. If that confidence rests on the explanatory power of sociological analysis, then it rests on a false foundation, for the sociologist has no explanation, it seems, of the success of science. But since rejection of the realist framework does not depend upon endorsement of a radical, sociological answer, perhaps Kuhn's confidence might find grounds elsewhere. Larry Laudan, for example, is, like Kuhn, no devotee of realism, yet takes himself to be in possession of an alternative explanation of scientific progress.[65] He holds that the success of science consists in problem-solving and that this success can be explained by the development of problem-solving techniques. Thus if this account were persuasive, it would provide Kuhn with a middle way between the realist position, which he rejects, and the sociological analysis, which is inadequate to the task.

The flaws in this bid to explain scientific success are exposed in Newton-Smith's criticism of Laudan's conception of science as a problem-solving venture.[66] Newton-Smith's charge is that by taking an 'agnostic' or 'atheistic' position on truth one loses the very right to talk of progress, let alone the ability to account for it.

---

[64] R. Boyd, 'The Current Status of Scientific Realism', in *Scientific Realism*, ed. J. Leplin (Berkeley: University of California Press, 1984), 60. And see W. H. Newton-Smith, 'The Role of Interests in Science' in *Philosophy and Practice* (Royal Institute of Philosophy Lectures, vol. 18, ed. A. Phillips Griffiths, Cambridge: Cambridge University Press, 1985), 59–73.

[65] L. Laudan, *Progress and its Problems* (London: Routledge & Kegan Paul, 1977). It should be noted that Laudan's doubts about truth are not the same as Kuhn's. The latter tends to find it conceptually troubling, the former epistemologically so: see Chapter 3 below. However this makes no difference to the question at issue here, which is whether it is possible to give an account of scientific progress and rationality which eschews reference to truth or verisimilitude.

[66] Newton-Smith, *Rationality of Science*, 185–92.

Hence he contradicts Kuhn's claim that nothing 'is lost, least of all the ability to explain scientific progress, by taking this position'.

For Laudan progress in science consists in increasing the problem-solving effectiveness of a research programme, this progress not requiring assessment of the truth or verisimilitude of the theory in question. He expounds his position as follows:

Rationality, it is usually argued, amounts to accepting those statements about the world which we have good reason for believing to be true. Progress, in its turn, is usually seen as a successive attainment of the truth by a process of approximation and self-correction. I want to turn the usual view on its head by making rationality parasitic upon progressiveness. *To make rational choices is, on this view, to make choices which are progressive* (i.e., which increase the problem-solving effectiveness of the theories we accept). By thus linking rationality to progressiveness I am suggesting that we can have a theory of rationality *without presupposing anything about the veracity or verisimilitude of the theories* we judge to be rational or irrational.[67]

Laudan's thesis relies on the claim that 'one need not, and scientists generally do not, consider matters of truth and falsity when determining whether a theory does or does not solve a particular empirical problem.'[68] But can science progress towards the goal of increased problem-solving effectiveness without reference to truth? Addressing this question, Newton-Smith bids us imagine a pair of theories T_1 and T_2, such that if a statement of a problem is entailed by T_1 its negation is entailed by T_2. Thus, imagine that T_1 implies the problem statement 'sugar dissolves in hot water', whereas T_2 implies the problem statement 'sugar does not dissolve in hot water'. Suppose further, as is natural, that we are inclined to believe the consequences of T_1 and disbelieve the consequences of its rival. That preference can be explained by saying either that T_2 solves only spurious problems (that is, its problem statements are false), or by saying that T_2 generates anomalies (that is, false predictions), but not by the simple claim that T_1 solves more problems than does T_2.

To take seriously the stricture against assessing a theory's consequences as true or false will result in a portrayal of scientific practice which is a parody. We would have to conclude that science

[67] *Progress and its Problems*, 125.
[68] Ibid. 24.

progresses as much in solving problems such as why sugar never dissolves in hot water, or why all swans are green, as it does in solving what we take to be genuine problems.[69] Avoiding this parody of science requires that we assess problems as genuine or spurious and their problem statements as true or false. Since scientific progress could consist in solving only real problems, not spurious, false ones, Laudan's goal of accounting for progress requires reference to truth or falsity. With truth excluded from our armoury of explanatory resources, we would be at a loss to specify a goal for science which reflects its practice. Hence making rationality parasitic on progress does not avoid concepts of truth and falsity, but depends upon them. Progress in problem-solving requires judgements of truth, in which case it is difficult to see what scruples we should have about describing theories themselves as true or false.

Laudan may concede so much to Newton-Smith's argument, yet reply that a significant form of epistemological instrumentalism may still obtain. Such instrumentalism would admit that there is, in Newton-Smith's words, 'a range of sentences which are such that we can have reasonable beliefs (in principle at least) about their truth value',[70] but would insist that this range includes only observational and not theoretical sentences. In other words, it would maintain that in judging a theory to be better at solving problems than a rival, we assess not the theory as such for truth or falsity, but only its observational consequences.

This reply is not one which Kuhn would make, for it relies on the observation/theory dichotomy which, as I have already noted, he rejects. But that is not the main point. Rather Laudan's imagined response is unsatisfactory because the observation/theory dichotomy which it would presuppose founders on an inability to provide a principled distinction of predicates into these two categories.[71]

It should not be denied that a distinction between observational and theoretical sentences is intuitively plain in many cases. The

[69] Newton-Smith, *Rationality of Science*, 187.
[70] Ibid. 190.
[71] The classical paper on this subject is G. Maxwell's 'The Ontological Status of Theoretical Entities', in *Minnesota Studies in the Philosophy of Science*, iii, ed. H. Feigl and G. Maxwell (Minneapolis: University of Minnesota Press, 1962), 3–27. See also: P. Achinstein, *Concepts of Science* (Baltimore: Johns Hopkins University Press, 1968); M. Hesse, *The Structure of Scientific Inference* (London: Macmillan, 1974); and Newton-Smith, *Rationality of Science*, 22–8.

class of theoretical sentences is made up of those which include terms such as 'electron', 'field', and 'charge'. Observational or non-theoretical sentences employ terms like 'red', 'warm', 'floats', 'hard', and so on. What, though, is the basis for the distinction? According to Achinstein, the usual answer is the one given by the Logical Positivists, that 'the distinction depends on the criterion of observability: terms on the theoretical list are supposed to refer to unobservables, those on the nontheoretical list to observables.'[72]

But if this is the basis of the distinction, why should it be accorded the epistemological significance which instrumentalists attribute to it? Bas van Fraassen's defence of what he terms 'constructive empiricism' in *The Scientific Image* hinges on the distinction between observable and unobservable, but his critics have latched on to his failure to demonstrate its significance.[73] Thus P. M. Churchland asks that we consider the various reasons why we fail to observe certain entities. In the first place he notes that their being unobserved may be explained by the fact that 'relative to our natural sensory apparatus, they fail to enjoy an appropriate spatial or temporal "position".' They may exist in a distant place or in a distant time. But there are, he continues, many other reasons why entities or processes may be unobserved:

they [may] fail to enjoy the appropriate spatial or temporal *dimensions*. They may be too small or too brief or too large or too protracted. Third, they may fail to enjoy the appropriate *energy*, being too feeble or too powerful to permit useful discrimination. Fourth and fifth, they may fail to have an appropriate *wavelength* or an appropriate *mass*. Sixth, they may fail to 'feel' the relevant fundamental forces our sensory apparatus exploits, as with our inability to observe the background neutrino flux, despite the fact that its energy density exceeds that of light itself.[74]

Churchland's point is that spatial or temporal distance is but one reason amongst many which render something beyond the bounds of human observation. Now there may be a practical point in calling something 'observ*able*' if it fails only the first test of spatio-temporal proximity and '*un*observable' if it fails any of the others, in that we naturally have some control over the spatio-temporal

[72] *Concepts of Science*, 158–9.
[73] Bas van Fraassen, *The Scientific Image* (Oxford: Clarendon Press, 1980).
[74] P. M. Churchland, 'The Ontological Status of Observables: In Praise of Superempirical Virtues', in *Images of Science*, ed. P. M. Churchland and C. A. Hooker (Chicago: University of Chicago Press, 1985), 39.

perspective. But this contingent fact can surely have no epistemological significance. That planets in the Andromeda galaxy are in principle observable in this sense, whereas electrons are not, does not make beliefs we have about the former any more secure. Churchland drives the point home with the comment:

> Had we been less mobile than we are—rooted to the earth like Douglas firs, say—yet been more voluntarily plastic in our sensory constitution, the distinction between the 'merely unobserved' and the 'downright unobservable' would have been very differently drawn. It may help to imagine here a suitably rooted arboreal philosopher named (what else?) Douglas van Fiirrsen, who, in his sedentary wisdom, urges an antirealist skepticism concerning the very *distant* entities postulated by his fellow trees.[75]

There is, it seems, no significant epistemological difference to be drawn between the observable and the unobservable. If that is the case, then instrumentalism which relies on the observation/theory dichotomy, and has this distinction as its basis, is shown to be poorly founded. Hence Newton-Smith's argument for the necessity of reliance on the categories of true and false in judging progress in problem-solving cannot be limited by the claim that such judgements need relate only to one class of sentences in science. For no principle has been provided which would justify such discrimination.

Conclusion

The purpose of this chapter has been in the first instance negative, though in later chapters I shall draw on some of the undoubted insights in Kuhn's work in developing a conception of the nature of scientific rationality. It has been my aim to highlight the inadequacies of a particularly important and influential understanding of science, one which poses a radical challenge to many commonly held assumptions about it. And I have been concerned to stress that a radical position proves unable to explain the success of science, even if it can, with important modifications, explain the formation of consensus amongst scientists. Neither appeal to sociology nor appeal to the notion of science as progressing through the refinement of problem-solving techniques helps. The one has nothing to contribute to the explanation of scientific success (even if

[75] 'Ontological Status of Observables', 39–40.

it has a role in explaining aspects of scientific change), and the other turns out to be reliant on the realist concepts it pretends to eschew.

Does this not suggest, then, that realism ought to be considered a candidate as an explanation of the nature of science in spite of Kuhn's worries? It is to the clarification and criticisms of realism that the next chapter is devoted.

3.

In Defence of Rational Realism

1. The Argument for Rational Realism

A PARTICULAR failing of Kuhn's philosophy of science, so I have claimed, is its inability to account for science's success. Yet this is one of science's most noteworthy features, and one which requires explanation. Thus Kuhn's failure in this regard leads us to examine with a fresh interest the claims of realism, which Kuhn put to one side, stressing the obscurity of the very concept of truth.

'Realism' is a slippery term, and the currently favoured position I wish to elucidate I designate 'rational realism'. Rational realism is to be distinguished from a position which we might label 'bare realism'. To be a bare realist about science is to assert no more than that the sentences of a theory are true or false in virtue of how the world is independently of ourselves. Asserting this—what Newton-Smith calls the 'ontological ingredient in realism'[1]—is not in itself enough to make one a rational realist, for it leads to no solution to the problem of justifying one theory as against another. For though one may allow that the sentences of scientific theories are true or false as the world dictates, there is as yet no commitment to the further claim that we are sometimes in possession of good reasons for saying of some of these sentences that they are either true or false. Maybe we can say of a theory that it is empirically adequate or inadequate (that it fits the data), but no more than that.

Now it is unlikely that the bare realist would claim the label 'realist', since bare realism will be somewhat esoteric and would play no part in an account of the progress of science, nor in an account of why we prefer science to, let us say, witchcraft. Some, of course, who describe themselves as 'realists' seem to their critics to be bare realists—so it might be said of Popper, if, as is often claimed, his repudiation of induction leaves him without grounds for holding that science's methods enable it to capture theories of

Rationality of Science, 29 f.

increasing verisimilitude. But if the bare realist acknowledges the implications of his or her position with its epistemological doubts, then it is likely that he or she will altogether eschew the term 'realism', even while rejecting an instrumentalist construal of the meaning of scientific theories. This is van Fraassen's stance in *The Scientific Image*.

The rational realist adds to this minimal, ontological realism an epistemological ingredient: a belief that we sometimes have good reason for concluding of two rival theories that one is more approximately true than another. Hereafter by 'realism' or 'rational realism' I mean to refer to the combination of the two elements, ontological and epistemological, which constitutes an interesting form of realism.

However else rational realists might argue their case, in recent years a certain argument for realism has won much favour. Though its advocates differ over subtleties, the argument is straightforward and proceeds from a premiss undisputed by all serious philosophies of science—that is that present-day science and technology are notably successful in enabling us to manipulate and cope with the world. It has been dubbed the 'no miracles argument', for realism's claim is that it is the only philosophy that does not make the success of science a miracle.[2]

The realist suggests that the success of science provides a warrant for science's truth or verisimilitude, and hence for the rationality of its methods. That is to say, the manipulative and predictive power offered by present-day science carries over to scientific theories a presumption as to their truth or verisimilitude. The burden of the argument rests on an alleged inference to the best explanation: for what other than the truth or verisimilitude of scientific theories, would explain the power offered by science, as against that to be

---

[2] The argument can be found in several authors in several forms. The most sustained defence of realism is W. H. Newton-Smith's *Rationality of Science*. Another advocate of this type of argument is R. Boyd; see his 'Realism, Underdetermination, and a Causal Theory of Evidence', *Noûs*, 7 (1973), 1–12; 'The Current Status of Scientific Realism'; and 'Lex Orandi est Lex Credendi', in *Images of Science*, ed. Churchland and Hooker, 3–34. The first of Boyd's articles is credited by H. Putnam as inspiration for his own defence of realism, in Lecture II of *Meaning and the Moral Sciences* (London: Routledge & Kegan Paul, 1978). E. McMullin's case for realism stresses the role of metaphor in science and the greater force of the realist argument in relation to sciences such as biology and geology over against physics—'A Case for Scientific Realism', in *Scientific Realism*, ed. Leplin, 8–40.

derived from tossing coins, practising witchcraft, or believing Greek myths?

Of inference to the best explanation Gilbert Harman writes:

In making this inference one infers, from the fact that a certain hypothesis would explain the evidence, to the truth of that hypothesis. In general, there will be several hypotheses which might explain the evidence, so one must be able to reject all such alternative hypotheses before one is warranted in making the inference. Thus one infers, from the premise that a given hypothesis would provide a 'better' explanation for the evidence than would any other hypothesis, to the conclusion that the given hypothesis is true.[3]

In the particular inference put forward by the rational realist it is alleged that the success of science would remain a mystery, a coincidence, or a miracle, unless we can answer the question which we feel compelled to ask: 'Why do these theories and these methods enable us to cope with the world?' The answer is that the theories enable us to cope because they are true or approximately true.

There are, of course, those who deny that realism is the way to dispel the mystery, for it is not the right explanation of science's success. But there are others who think that the sense of mystery is itself misplaced and that there is nothing at all mysterious in the success of science. Van Fraassen, in *The Scientific Image*, offers three related objections which seek to threaten the initial force of the argument sketched above.

The realist argument is founded on an inference to the best explanation, but in his first objection van Fraassen alleges that the success of science does not need this explanation, for the fact that our theories get better is not surprising. He writes that

the success of current scientific theories is no miracle. It is not even surprising to the scientific (Darwinist) mind. For any scientific theory is born into a life of fierce competition, a jungle red in tooth and claw. Only the successful theories survive—the ones which *in fact* latched on to the actual regularities in nature.[4]

Darwinism bids us ask not why mice run away from cats, but rather

[3] G. Harman, 'The Inference to the Best Explanation', *Philosophical Review*, 74 (1965), 89. In his discussion of inference to the best explanation, Harman notes that what he puts under this label Peirce termed 'abduction', and others have called 'the method of hypothesis', 'hypothetic inference', 'the method of elimination', 'eliminative induction' and 'theoretical inference'.

[4] *Scientific Image*, 40.

note that ones which do not, cease to exist. In the same way, it is hardly surprising that theories chosen for their ability to cope with the world are part of a series of ever more successful theories. What the common-sense argument for scientific realism says needs explaining is perfectly well explained already.

This objection to the rational realist's argument merely misses the mark. Whilst it might be true that mice which do not run away tend not to survive, there is a perfectly reasonable question left unanswered by the Darwinian observation—that is, what it is about a mouse which causes it to run away. The same question will arise in relation to successful theories. Although it is only theories which cope which will survive, it is still legitimate to consider in virtue of what they are able to cope. Musgrave asks rhetorically: 'Do we explain why some theory is empirically adequate . . . by remarking that theories which are not have been eliminated?'[5]

Van Fraassen has a second criticism of the argument from the success of science. He thinks that it is guilty of pushing explanation to excess and, in effect, appeals to something like the principle of sufficient reason. Yet such a principle is fundamentally at odds with the predominant interpretation of quantum physics, which rejects the search for hidden variables so as to render the discipline deterministic.[6]

Once again this objection comes to nothing, for the realist is not committed to saying that there *must* be an explanation of every observed regularity. On the contrary the realist's position is well expressed in Devitt's remark: 'Though it is not obvious where explanation should stop, the best evidence that it has not gone far enough is a good explanation that goes further.'[7] It is not that there must be an explanation, but that, where one is proffered, it is to be judged on its merits. Thus the realist's attempt to explain the observed regularity in question, namely the success of science, does not rely on a mistaken, old-fashioned or a-priori demand for explanation and is free from the taint of the principle of sufficient reason.

There is a third preliminary objection to the realist's argument,

[5] A. Musgrave, 'Constructive Empiricism *Versus* Scientific Realism', *Philosophical Quarterly*, 32 (1982), 266. The same point is made in Newton-Smith's 'Realism and Inference to the Best Explanation' (unpublished paper, 1984).

[6] *Scientific Image*, 23–34.

[7] M. Devitt, *Realism and Truth* (Oxford: Blackwell, 1984), 127.

inspired by van Fraassen, which it will be appropriate to mention here. It arises from the reasonably uncontroversial point that if explanations are in a sense answers to 'why-questions', then they are, as Lucas puts it, 'to some extent *ad hominem*'.[8] Putting the same point rather more formally van Fraassen writes:

to say that a given theory can be used to explain a certain fact, is always elliptic for: there is a proposition which is a telling answer, relative to this theory, to the request for information about certain facts (those counted as relevant for *this* question) that bear on a comparison between this fact which is the case, and certain (contextually specified) alternatives which are not the case.[9]

Van Fraassen means to draw attention to the point that in giving an answer to a question one gives the one most appropriate to the doubts of the questioner. One of the examples he discusses is the question 'Why did Adam eat the apple?'[10] The enquirer may be wondering, say, why it was Adam (and not Eve) who ate the apple. Or perhaps why it was the apple which Adam ate, as opposed to any other fruit he might have found in the garden; or why Adam ate the apple rather than returning it to Eve. The different questions implicit in the one question demand different answers. 'Because he was hungry' may be a good reply in one case and not in another.[11]

It has been suggested that the relativity of explanation threatens the realist case. How can explanatory inferences carry truth if it is allowed that there may be different explanations of the same phenomenon? But though van Fraassen somehow connects the contextual nature of explanations with his attack on scientific realism, and Putnam is impressed as to its significance, in fact nothing deep seems to follow from it.[12] Of course we select which explanation to give according to the questioner's supposed interests; but this relativity of explanations is not incompatible with

---

[8] J. R. Lucas, *The Freedom of the Will* (Oxford: Clarendon Press, 1970), 35.

[9] *Scientific Image*, 156.

[10] Ibid. 127.

[11] The fact that responses to requests for explanation are relative to interests is the basis for certain well-worn children's jokes, such as the one which asks 'why did the elephant cross the road?' and replies, 'to get to the other side'. The effect of the joke, if there is one, depends upon the pragmatic nature of explanation; this was not the answer that we expected.

[12] For Putnam's argument see *Meaning and the Moral Sciences*, 41–5, and Newton-Smith's discussion of it in 'The Role of Interests in Science', 64–8.

the notion of there being a true explanation. The question put to a robber: 'Why do you rob banks?' will receive a different answer when put by a fellow robber than when asked by a priest—but although the partial explanations proffered will be relative to the questioner's interests, they are not competitors and so both may be true (though neither is the whole truth). So too in the case of the success of science, different explanations may be offered which are not in competition with realism. The question why such and such a branch of science produced so many useful technological applications will be answered in certain contexts by pointing to the amount of money which was spent on research and development, without casting doubt on an explanation in terms of the verisimilitude of the underlying theories which the technology presupposes.

Having dealt with those who think that the realist's question is naïve, does not need answering, or rests on a mistaken concept of explanation, I turn to the suggestion that it is not the question, but rather the realist's answer, which is wrong. It will be recalled that rational realism has two elements, epistemological and ontological, so that it can be undermined by attacks on either one. The objections to realism which I shall consider may, then, be divided according to their aim, the first attacking the ontological leg of rational realism, the second the epistemological one.

According to the first objection it is mere metaphysical sophistry to talk of our theories being true in virtue of how the world is. The notion of a correspondence between theories and the world is fundamentally suspect. The second objection makes a different point, and comes from those who do not question the conceptual propriety of the proposed explanation, but who insist that study of the history of science rules out the claim that this is a *good*, as opposed to a *possible*, explanation of scientific success.

The first objection is an objection to bare realism, since it attacks the ontological element in realism. It stems from doubts about truth such as those which we have already found in Kuhn's work and which are more fully developed by Richard Rorty in arguments to be considered in the next section. The second objection is encouraged by Kuhn's emphasis on the history of science and is pressed forcefully in the writings of Larry Laudan, which will be considered in section 3 of this chapter.

2. *Wittgenstein, Rorty, Davidson, and the Sense of Truth*

In his *On Certainty* Wittgenstein expresses doubts about certain uses of the concept of truth. He asks:

Well, if everything speaks for a hypothesis and nothing against it—is it then certainly true? One may designate it as such.—But does it certainly agree with reality, with the facts?—With this question you are already going round in a circle.[13]

And he puts this question to the realist:

What does this agreement consist in, if not in the fact that what is evidence in these language games speaks for our proposition? (*Tractatus Logico-Philosophicus*)[14]

The reference to the *Tractatus Logico-Philosophicus* serves to remind us of the central contrast between *On Certainty* and that early work, in which it is maintained that the possibility of making sense of agreement with reality is an a-priori requirement of language. In the *Tractatus* a proposition derives its sense from the possibility of this agreement with reality.[15] The agreement or disagreement of this sense with reality constitutes a proposition's truth value.[16] Thus tautologies and contradictions which respectively rule out nothing or everything, and hence do not 'picture', are held to be degenerate propositions, 'senseless'. But Wittgenstein's later judgement on this project, and on those realist projects which follow in its footsteps, is that 'the idea of "agreement with reality" does not have any clear application.'[17]

As we have seen in the previous chapter, Kuhn has similar doubts about truth: 'the notion of a match between the ontology of a theory and its "real" counterpart in nature now seems to me illusive in principle.'[18] But in spite of these doubts neither Kuhn nor Wittgenstein, with their worries about truth, would wish to be regarded as philosophical sceptics. They hold not that our theories fail to achieve an accolade which other theories might achieve, but

---

[13] *On Certainty*, para. 191.
[14] Ibid. para. 203.
[15] L. Wittgenstein, *Tractatus Logico-Philosophicus*, trans. D. F. Pears and B. F. McGuinness (London: Routledge & Kegan Paul, 1961), 4.2.
[16] Ibid. 2.222.
[17] *On Certainty*, para. 215.
[18] Postscript, 206.

that the accolade of truth is not sensibly granted to any theory; that is, that the concept of truth or verisimilitude is senseless when applied to scientific theories taken as a whole. As Kuhn puts it, ' "truth" may, like "proof", be a term with only intra-theoretic applications.'[19]

Now if the project of 'grounding' science is illusory because it rests on some conceptual confusion, then it would be unfair to accuse realism's opponents of scepticism, just as it would be unfair to describe as a sceptic one who does not believe in round squares. And if these critics are right, the argument for realism would founder in that its central explanatory concept would be shown to be incoherent in this context.

A present-day spokesman for this position, which he terms 'pragmatism' and attributes to both Wittgenstein and Kuhn, is Richard Rorty. According to the pragmatist, both realism and scepticism rest on a grand mistake:

we . . . counter Kant by saying that 'Only the pragmatist can be an empirical realist'—that is, only if we give up the notion of legitimation can we rest content with accepting the knowledge-claims of science at face value, since no skeptical attack, and no anti-skeptical project of legitimation such as Kant's, can succeed.[20]

Rorty tells us that the pragmatist is best conceived of as the most radical of anti-Platonists, and should be contrasted with the positivists and their half-hearted anti-Platonism. For whereas the positivist is inclined to attack the notion of truth as it is applied to all areas, save that of science, the pragmatist 'drops the notion of truth as correspondence with reality altogether, and says that modern science does not enable us to cope because it corresponds, it just plain enables us to cope.'[21]

Having been assured that the expectation of grounding world

[19] 'Reflections on my Critics', 266.

[20] R. Rorty, 'Transcendental Arguments, Self-Reference and Pragmatism', in *Transcendental Arguments and Science*, ed. P. Bieri, R. Horstmann and L. Krüger (Dordrecht: Reidel, 1979), 84. A claim similar to Rorty's that pragmatism amounts to 'accepting the knowledge-claims of science at face value' is found in Arthur Fine's 'The Natural Ontological Attitude', in *Scientific Realism*, ed. Leplin, 83–107. In 'A Case for Scientific Realism' (26) McMullin notes that this assertion gives the position 'a puzzling sort of undeclared status where [its advocates] appear to have the best of both worlds'. He continues: 'I am inclined to think that their effort to have it both ways must in the end fail.'

[21] R. Rorty, *The Consequences of Pragmatism* (Brighton: Harvester, 1982), p. xvii.

pictures is a philosophical illusion, we may pass on with a clear conscience to the tasks which Rorty sets out for 'a deconstructed philosophy', a philosophy rid of its self-image as overseer of our culture, whose practitioners would be content to function as edifying 'name-droppers'.[22] This new philosophy stands over against Philosophy (the capital signifying the subject as it is traditionally conceived, stemming from Descartes's search for secure foundations), and is edifying not constructive. It could even be termed 'satire', as in Rorty's characterization of *Philosophical Investigations*.[23] Its principal aim is to rid us of a picture, and the problems which that picture generates, not to propose a new picture. Its form will be 'don't ask these questions any more', and it employs suggestion not demonstration. Thus Rorty writes: 'If one thinks of the end of philosophy in these terms, it is quite clear that it is not the sort of thing which might be brought about by exposing some confusions, or marking the boundaries of areas of discourse, or pointing out some "facts about language".'[24]

So why should the realist give up the project of 'legitimation', the project of justifying the claims of science? What exactly does this attack on the ontological ingredient in realism amount to? For the pragmatist does not have an argument against realism, if by an argument one understands some demonstration of conceptual impropriety. Rorty's proposal that we drop the notion of truth as correspondence rests rather on the contention

that several hundred years of effort have failed to make interesting sense of the notion of 'correspondence' (either of thoughts to things or of words to things). The pragmatist takes the moral of this discouraging history to be that 'true sentences work because they correspond to the way things are' is no more illuminating than 'it is right because it fulfills the Moral Law'. Both remarks, in the pragmatist's eyes, are empty metaphysical compliments—harmless as rhetorical pats on the back to the successful inquirer or agent, but troublesome if taken seriously and 'clarified' philosophically.[25]

The realist will wonder what it is for a concept 'to make

[22] *Consequences of Pragmatism*, p. xl; and for a fuller vision of the new philosophy, see R. Rorty, *Philosophy and the Mirror of Nature* (Oxford: Blackwell, 1980), chs. 7 and 8.

[23] R. Rorty, 'Keeping Philosophy Pure: An Essay on Wittgenstein', in *Consequences of Pragmatism*, 34.

[24] Ibid. 33. In this essay Rorty had not adopted the practice of the pejorative capital for 'Philosophy'.

[25] *Consequences of Pragmatism*, p. xvii.

interesting sense' and may well insist, along with McMullin, that a concept makes interesting sense if it plays a key role in an argument, and is not demonstrably confused or incoherent. A particular analysis of 'truth' or 'verisimilitude' may be found wanting, but the realist, whilst admitting this, may yet contend in reply that the mere want of a satisfactory analysis should not cause too much concern. If appeal to truth has explanatory power as the realist claims, then we should no more worry ourselves about its use than scientists bother themselves about the use of the concept 'gene' prior to a full description of what genes are.[26] If one is going to assert the senselessness of truth, then one needs to produce more than a catalogue of failed projects of elucidation or definition. One needs to demonstrate that the realist's use of the concept is incoherent.[27]

The realist, then, does not find Rorty's proposal for a new agenda for philosophy particularly compelling. The realist's claim is that an account of the success of science, which includes truth or verisimilitude as its central explanatory concept, does make interesting sense and that nothing less than an a-priori demonstration of a conceptual muddle involved in realism would give grounds for rejecting realism at this early stage.

This, however, does not acquit realism on all the charges Rorty lays against it. For whilst Rorty usually presents pragmatism as the dropping of certain intractable questions, at other times he presses its case more forcefully, as if there has been a demonstration of the confusions upon which realism relies. Thus on occasions Rorty implies that entertaining the question as to the truth or falsity of world pictures is not so much futile as demonstrably conceptually confused or incoherent, and that he possesses an argument which would give substance to the claims made by Kuhn and Wittgenstein that 'truth' is a concept which makes sense only when applied within, not between, theories. In other words, Rorty claims possession of an argument which goes beyond the assertion that rational realism does not make 'interesting sense' and seeks to demonstrate realism's senselessness.

The claim to possess such a powerful and decisive argument is

---

[26] A point made by Newton-Smith, *Rationality of Science*, 197.

[27] The apologist for traditional religious faith will be familiar with the ploy, found, for example, in attacks on two-nature Christology, of confusing the admitted inadequacy of many explications of a concept with an admission of that concept's incoherence.

found in Rorty's article 'Transcendental Arguments, Self-Reference and Pragmatism', in which he contends that Donald Davidson's 'On the Very Idea of a Conceptual Scheme' works 'against the whole problematic of legitimation created by the scheme–content distinction.'[28] Even here Rorty's claim is not unequivocal: we are told that, though one would like an argument for pragmatism, the closest one can get to this is a series of *ad hominem* rebuttals of the various attempts to make sense of correspondence.[29] But despite this, Davidson's paper is somehow thought to be different and more effective:

the aim of his argument is to make impossible the whole Cartesian and Kantian dialectic which makes skepticism and anti-skeptical transcendental argumentation possible. I construe (*pace* its author) Davidson's argument against the notion of 'conceptual scheme' and against the 'scheme–content distinction' as an argument for pragmatism, and thus against the possibility of epistemology. Davidson, in other words, seems to me to have found a transcendental argument to end all transcendental arguments—one which tears down the scaffolding upon which the standard paradigms of 'realistic' transcendental arguments were mounted.[30]

Thus Rorty claims to possess an argument which demonstrates a conceptual confusion in realism, namely, its reliance on the illegitimate notion of scheme and content. That argument is thought to depend on Davidson's paper, which must be examined if we are to see whether or not it warrants the implications which Rorty draws from it.

Davidson's concern is with the relativism which bases itself on the notion of conceptual schemes organizing our experience, and which suggests that:

There may be no translating from one scheme to another, in which case the beliefs, desires, hopes and bits of knowledge that characterize one person have no true counterparts for the subscriber to another scheme. Reality

[28] 'Transcendental Arguments, Self-Reference and Pragmatism', 99. D. Davidson, 'On the Very Idea of a Conceptual Scheme', *Proceedings of the American Philosophical Association*, 47 (1973–4), 5–20. For useful discussions of Davidson's paper see T. R. Mills, 'Relativism in the Analysis of Religious Belief and Language and the Construction of Theories of Meaning for Natural Languages', D.Phil. thesis (Oxford, 1982), ch. 9, and W. Carl's 'Comment on Rorty', in *Transcendental Arguments and Science*, ed. Bieri *et al.*, 105–12.
[29] 'Transcendental Arguments, Self-Reference and Pragmatism', 90.
[30] Ibid. 78.

itself is relative to a scheme: what counts as real in one system may not in another.[31]

Davidson holds that under examination this exciting form of relativism turns out to be unintelligible; we are unable to make sense of two radically different conceptual schemes or of the scheme–content distinction employed in describing such schemes. It is from this conclusion that Rorty draws the impossibility of epistemology.

Davidson's argument can be set out in the following series of steps:

1. An inability to translate between two languages (in whole or in part) is a necessary condition for their possessing different conceptual schemes. Davidson writes:

We may accept the doctrine that associates having a language with having a conceptual scheme. The relation may be supposed to be this: where conceptual schemes differ, so do languages. But speakers of different languages may share a conceptual scheme provided there is a way of translating one language into the other.[32]

If translation is possible, then there may be differences of belief between speakers, but not differences of conceptual scheme. For the exciting relativism we are considering argues for 'dramatic incomparability'—Davidson notes that ' "Incommensurable" is . . . Kuhn and Feyerabend's word for "not intertranslatable".'[33]

2. A second condition on there being different conceptual schemes is that the two languages said to contain them should recognize each other as languages. This suggests to Davidson a 'short line' with the issue: 'nothing, it may be said, could count as evidence that some form of activity could not be interpreted in our language that was not at the same time evidence that that form of activity was not speech behavior.'[34] Davidson's point is that in attributing a language to someone, one imputes a whole range of attitudes 'such as belief, desire and intention'. If we were unable to translate the supposed language, we would surely be on less than solid ground in claiming that the 'speakers' possessed such attitudes. What could it be which suggested that they were not just making a range of noises?

[31] Ibid. 5.   [33] Ibid. 12.
[32] Ibid. 6.   [34] Ibid. 7.

Davidson does not leave the issue here though:

There can be no doubt that the relation between being able to translate someone's language and being able to describe his attitudes is very close. Still, until we can say more about *what* this relation is, the case against untranslatable languages remains obscure.[35]

3. Davidson therefore puts on one side the 'short way' and tries another tack. The idea of an untranslatable language relies on 'the dualism of scheme and content, of organizing system and something waiting to be organized'. For, Davidson tells us, 'the common relation to experience or the evidence is what is supposed to help us make sense of the claim that it is languages or schemes that are under consideration when translation fails.'[36]

4. Now for a sentence or theory to fit our sensory promptings, face the tribunal of experience, or whatever the metaphor of scheme and content suggests, amounts to its being 'borne out by the evidence', where 'evidence' is all possible evidence.[37] But, so Davidson claims, 'the notion of fitting the totality of experience, like the notions of fitting the facts, or being true to the facts, adds nothing intelligible to the simple concept of being true.'[38]

5. The issue, therefore, comes down to what we can make of true but untranslatable languages or schemes.[39]

6. And since Davidson thinks that we do not understand the notion of truth applied to language apart from the notion of translation, we have arrived at an impossible criterion of a conceptual scheme different from our own. He writes:

according to Tarski's convention T, a satisfactory theory of truth for a language L must entail, for every sentence *s* of L, a theorem of the form '*s* is true if and only if *p*' where '*s*' is replaced by a description of *s* and '*p*' by *s* itself if L is English, and by a translation of *s* into English if L is not English. This isn't, of course, a definition of truth, and it doesn't hint that there is a single definition or theory that applies to languages generally. Nevertheless, Convention T suggests, though it cannot state, an important feature common to all the specialized concepts of truth. It succeeds in doing this by making essential use of the notion of translation into a language we know. Since Convention T embodies our best intuitions as to how the concept of truth is used, there does not seem to be much hope for a test that a conceptual scheme is radically different from our if that test depends on the assumption that we divorce the notion of truth from that of translation.[40]

---

[35] 'Transcendental Arguments, Self-Reference and Pragmatism', 8.
[36] Ibid. 12.   [37] Ibid. 15.   [38] Ibid.16.
[39] Ibid.   [40] Ibid. 17.

Davidson claims to have isolated a difficulty for the advocate of conceptual relativism—namely that conceptual relativism involves the assumption that theories might be true yet untranslatable when our 'best intuitions' about the use of the concept of truth make essential use of the notion of translation into a known language. The incoherence involved in relativism concerning truth can be put another way. For the conceptual relativist asks us to consider that something might be true in one scheme but false in another; to suppose that a sentence or proposition p is true in X but false in Y. But to suppose this is nonsense: if the meaning of a sentence is given by its truth conditions, then one could only suppose that p was true in X and false in Y by assuming that what was false in Y was not really a translation of what was true in X—for if there is a difference in truth value there must be a difference in truth conditions, and for there to be a difference in truth conditions is just what we take as evidence of a difference in meaning.[41]

7. Might we not, however, have a form of relativism which was based on the idea of partial, rather than complete, failure of translation? Against this Davidson argues that any theory of translation must make a start with a principle of charity; that is to say, that to give a translation involves attributing beliefs, intentions, and so on. But 'Since knowledge of beliefs comes only with the ability to interpret words, the only possibility at the start is to assume general agreement on beliefs.'[42] We shall have to begin by assuming that our subjects believe what we would believe if we found ourselves in their situation. So starting from a required assumption of charity, the suggestion that a different conceptual scheme is embodied in the other language cannot arise: 'Given the underlying methodology of interpretation, we could not be in a position to judge that others had concepts or beliefs radically different from our own.'[43]

Davidson ends his paper with the claim that 'we have found no intelligible basis on which it can be said that schemes are different. . . . if we cannot intelligibly say that schemes are different, neither can we intelligibly say that they are one.'[44]

The purpose in considering Davidson's argument was to assess Rorty's view that it works 'against the whole problematic of

---

[41] Newton-Smith, *Rationality of Science*, 35–37.
[42] 'The Very Idea', 18–19.
[43] Ibid. 20.
[44] Ibid.

legitimation created by the scheme–content distinction', and therefore as an argument for pragmatism. My concern, then, is not with Davidson's position as such, but rather with its alleged significance for Rorty's brand of pragmatism. The question to ask is whether any problem of legitimation remains in the face of Davidson's argument. In particular is the realist's project of legitimation redundant, and, in so far as it assumes the scheme–content distinction, senseless?

The contradiction which Davidson identified amounted to holding that there might be two languages or schemes which were both true but untranslatable. The exciting relativism which says that what is 'real' or 'true' in one scheme may not be so in another is thereby shown to be insupportable. But it is difficult to see why Rorty should suppose that this puts an end to epistemology. The rational realist's argument does not assume the possibility of relativism concerning truth, and is not addressed specifically to those who say that science is true for us whereas some other way of looking at the world may be true for others. Rather the realist's argument is directed against those who doubt that a particular scheme is true or approximately true. Just because we cannot make sense of two schemes both fitting the facts does not show that we cannot understand the possibility that one might fail to.

Of course the opponents of rational realism are unlikely to be using the notion of conceptual schemes in the strict sense defined by Davidson, and are more probably using it to highlight vast differences of belief. The claim is not, then, that two 'conceptual schemes' in this attentuated sense might be true, but that two conceptual schemes—to avoid confusion let us call them *Weltbilder*—might be taken for true, and that we presently have no way of adjudicating between them. Thus some talk of competing *Weltbilder* comes down to talk of differences of belief which arise as the gap between our limited range of evidence and our theories about the world are bridged by different methods of reasoning. It is the claim of rational realism that it provides an argument for accepting that form of bridging which is found in contemporary science.

Against Rorty's claims for Davidson's argument, I find nothing in it which rules out the sort of limited scepticism outlined in the last paragraph about what is taken for true. The principle of charity of translation of course shows that in interpreting a different *Weltbild* one cannot but set out by assuming widespread agreement in belief, including a common hold on the laws of logic and the like. But as I shall have cause to say in the next chapter, a plausible understanding of this principle of charity does not rule out that others will have beliefs different from our own—such as that the world is flat or that certain dances really do cause rain or the belief of a particular Brazilian Indian that he was a red parrot.[45] It merely rules out a radical difference of belief where there is no common ground at all.

Of course Davidson himself makes no claims for his argument of the sort made by Rorty, and he seems to have no intention of denying that there must be a project of legitimation, for the existence of differences of belief is just what gives the project a point. All that Davidson seeks to show is the incoherence of supposing that there might be two true conceptual schemes, not the incoherence of supposing that some of our beliefs may be wrong.

Rorty's hope for Davidson's argument was that when the very idea of a conceptual scheme was seen to be empty, the project of legitimation would, along with relativism, cease to be an issue. In fact that is to put the point too weakly: the argument would lead us to see that realism, scepticism, and epistemology in general involve conceptual confusions. But Davidson's argument cannot be employed to serve this end. Thus Rorty has no argument for pragmatism, beyond the weak suggestion that we just give up the search for 'truth' since it has in the past proved inconclusive and conceptually troubling. As yet there seems to be no demonstration that the concept of truth is somehow senseless, only an assumption that because its analysis has proved problematic it must be so. Rorty would like to be able to tell us that realism fails because it is incoherent, but ends up with the rather different position that realism is incoherent because it fails.

I conclude, then, that the attacks on the ontological ingredient in realism do not carry conviction. But the argument against rational

---

[45] For this last claim see T. E. Hulme's account, derived from Lévy-Bruhl, of a missionary convinced, despite all his efforts, that a Brazilian Indian really was claiming to be a red parrot. The Indian refused the missionary's suggestion that perhaps he meant that he would become a red parrot on his death, or rather that he was somehow related to a red parrot. As Hulme recounts the tale, 'The Indian rejected both these plausible attempts to explain away a perfectly simple fact, and repeated quite *coldly* that he *was* a red parrot.' T. E. Hulme, *Speculations*, 2nd edn., ed. H. Read (London: Kegan Paul, Trench & Trubner, 1936), 67.

realism has another element involving doubts about the ability of the realist to provide grounds for holding that one theory is, over against another, more approximately true. It is to these doubts that I now turn.

3. Is Realism the Best Explanation?

The first major objection to realism stems from doubts about truth, and has been considered in my discussion of Wittgenstein, Kuhn, Rorty, and Davidson. It alleged that the notion of our theories being true or false in virtue of how the world is, rests on some sort of conceptual confusion. The second objection to be studied is the one which says that confronted by the facts of scientific progress, realism cannot sustain its claim to be the best explanation.

The realist holds that the best explanation of the success of science is the truth or approximate truth of many of its theories. But is an inference to realism really an inference to the best explanation? In a paper entitled 'A Confutation of Convergent Realism' Larry Laudan attacks many of the realist's articles of faith.[46]

Laudan may be taken to be arguing on two main fronts, asking two questions of the realist. First, does the realist hypothesis that many current scientific theories are true or approximately true make likely what it claims to explain (that is, the success of science)? And second, even if the realist hypothesis does make likely what it seeks to explain, can the realist establish that this explanation is the best?

Laudan's answers to these questions give no comfort to the realist. On the first point he argues that approximate truth is not manifestly a sufficient condition for success, and since approximate truth has not been shown to imply success, it cannot be said to explain it. So realism fails as an explanation. Second, he argues that approximate truth is not a necessary condition for success. Even if, contrary to his first point, approximate truth is explanatory of success, it by no means follows that success is *best* explained by a supposition of approximate truth. In other words an inference to approximate truth from success is illicit, since a theory may be successful and yet not approximately true. Hence even if truth were

[46] L. Laudan, 'A Confutation of Convergent Realism', *Philosophy of Science*, 48 (1981), 19–49.

a possible explanation of the success of a scientific theory, it is by
no means the best.

Laudan isolates the main themes of epistemic realism in the
following theses:

(T1) if a theory is approximately true, then it will be explanatorily successful; and

(T2) if a theory is explanatorily successful, then it is
probably approximately true.[47]

The significance and form of (T1) is pressed by Laudan. The realist
argument moves from the power and success of science to a
presumption of its verisimilitude; that is to say, the realist inference
relies not on the premiss:

(a) true theories will be successful;

but rather on:

(b) theories possessed of verisimilitude will probably be successful; or, more accurately: if a theory T_2 is a better approximation to the truth than a theory T_1, then it is likely that T_2
will have greater predictive power than T_1.

Clearly (b) is intuitively weaker than (a), and the realist would
rather employ (a) than (b). That would, however, require an
argument for the truth of present-day science, which would be hard
to make in the face of the so-called 'pessimistic induction'. The
pessimistic induction notes that all the best theories of previous
epochs have turned out in due course to be strictly speaking false. It
is reasonable to suppose, then, that 200 years from now, let us say,
all the theories we presently favour will also be judged false. Hence
this induction precludes us from claiming that our present science is
true. At best we may assert that it is closer to the truth than were its
predecessors.[48]

Laudan admits, of course, that the link between the truth of a
theory and its success is compelling. If a theory is true, then any of
its consequences is true. However, Laudan asks whether the same
goes for verisimilitude. Why assume that if a theory is approximately true its consequences are more likely true than false?[49] He
warns that we should beware of a realist 'sleight of hand' which

[47] Ibid. 30.
[48] The pessimistic induction is to be found in Laudan's *Progress and its Problems*,
125 f.
[49] 'Confutation', 30–2.

trades on the obvious connection between truth and success to convince us of the separate thesis connecting approximate truth and success. Does verisimilitude entail and thereby explain success? Laudan maintains that, lacking a satisfactory analysis of verisimilitude, we would be wrong to say that it does: '*none* of the proponents of realism has yet articulated a coherent account of approximate truth which entails that approximately true theories will, across the range where we can test them, be successful predictors.'[50]

Laudan's objection would best be dealt with by provision of an analysis of verisimilitude, and Newton-Smith (dissatisfied with Popper's treatment of the problem and in anticipation of such an objection) tried to provide such an analysis showing how verisimilitude implies success.[51] He now acknowledges that his proposal suffers from profound difficulties, and that Laudan's demand for an analysis remains unanswered.

Does Laudan win this point then? Is the realist intuition about approximate truth implying success improper until such an analysis is provided? So Laudan claims, but the realist thinks that providing an analysis is not the only way to respond. The realist, applying once more an argument of Newton-Smith's used against Rorty, might say this: the alleged link between verisimilitude and success is intuitively plausible and explanatorily useful. The intuitive plausibility is clear enough. For imagine an explorer being given a choice between a map which is plain false and one which is described as possessing verisimilitude (that is, true in some respects or near to the truth). Naturally, the map which has verisimilitude would be chosen, and we would expect those who used it to find their way with more success than otherwise. Now suppose (as in this case) that the link between success and verisimilitude is explanatorily useful. Then the realist will insist that the mere fact that the concept is formally unanalysed does not show that it is improper to rely upon it, and upon its explanatory power. For if a concept has an explanatory role, its use may be legitimate even while its analysis remains problematic.

This pattern is observed and accepted in science. Concepts such as 'plates' (in geology), 'genes', 'attraction' (in Newton's theory), and 'field', were used while yet unanalysed. That is to say, whilst an

---

[50] 'Confutation', 32.
[51] For the details, see Newton-Smith, *Rationality of Science*, 52–9 and 198–207.

exact specification or definition of the terms was lacking, theories continued to be used in which these concepts connected up in various ways to do the explanatory work required of them. For the intuitive promise of the concepts, often understood metaphorically, was enough to warrant reliance on them whilst they were further investigated.

Of course it is the burden of Laudan's *Progress and its Problems*, as we have already seen in Chapter 2, that the explanatory task which is given to 'verisimilitude' is perfectly well performed without it. In other words, his argument was that scientific progress may be explained without reference to truth. Thus he shares Kuhn's view that although one disdains the realist's concept of truth, nothing 'is lost, least of all the ability to explain scientific progress, by taking this position'.[52] But against this it has been argued that the proposed separation of truth and progress cannot be achieved, for the notion of science as having progressed through the solving of problems requires in turn that we employ the concept of truth in assessing beliefs. Since the observation/theory dichotomy is to be rejected, there remains no principle by which we may object to thinking of theories (webs of belief) as true or false or—since the pessimistic induction forbids us from saying that theories are true *per se*—as nearer the truth. But the important point now is that Laudan's account of progress having failed, there is no obvious alternative to verisimilitude if we are to explain scientific progress. It may be that this explanation itself fails to be satisfactory, but prior to establishing that, the prima facie explanatory promise of the concept of verisimilitude warrants our reliance on it.

The assumption that an analysis of verisimilitude is still wanting is not in itself fatal to the realist case, and we may move on to Laudan's second objection. This objection is directed to the other principal realist thesis, namely:

(T2) if a theory is explanatorily successful, then it is probably approximately true.

Laudan argues that even allowing that if a theory is approximately true it will probably be successful, there is no reason for allowing an inference the other way; that is, for inferring that if a theory is successful it is probably approximately true. Success could be taken as a warrant for a judgement of approximate truth only if

---

[52] 'Reflections on my Critics', 265.

approximate truth were a necessary condition for success. For even if approximate truth is sometimes the explanation of success, it does not follow that it always is.[53]

The realist cannot maintain the propriety of the principle that if a theory is successful then it is probably true, so Laudan maintains, because realism also commits one to the view that for a theory to be approximately true its central terms must refer. Now since there are, Laudan suggests, a number of successful theories whose central terms do not, by present lights, refer, it must follow that not all successful theories are approximately true. Laudan lists these theories: the crystalline spheres of ancient and medieval astronomy; the humoral theory of medicine; the effluvial theory of static electricity; catastrophist geology, with its commitment to a universal (Noachian) deluge; the phlogiston theory of chemistry; the caloric theory of heat; the vibratory theory of heat; the vital force theories of physiology; the electromagnetic ether; the optical ether; the theory of circular inertia; theories of spontaneous generation. Laudan's conclusion is that 'the history of science offers us . . . a plethora of theories which were both successful and (so far as we can judge) non-referential'.[54] So, Laudan hopes, the realist is driven to admit the existence of a range of successful theories, the success of which cannot be accounted for by pointing to their approximate truth since their failure to refer excludes such a supposition. Thus the existence of these successful theories is a fatal flaw in the inference from success to approximate truth posing as an inference to the *best* explanation.

Laudan's second objection—that there are successful but untrue theories, which suggests that an inference from success to approximate truth is illegitimate—is open to a number of different answers. His case rests on the claim that there are many theories which although untrue because non-referring, are none the less successful. The claim can be broken down into several elements: that these theories are successful, that they are non-referring, and that because they are non-referring they cannot be approximately true. The realist might challenge one or all of the three elements in the argument.

To begin with, the realist may ask how many of the theories

[53] 'Confutation', 32–6.
[54] Ibid. 33.

listed are genuinely successful. 'Success' in the realist's vocabulary, Laudan protests, is not a sharply defined term,[55] which should in itself be a cause for caution. In Chapter 6 I shall consider an account of success which the realist might give. For the moment, however, I note that the success the realist principally means to highlight is observational success, an ability to account for known observations and to predict new ones, and that the realist understands this to be a stricter condition than mere compatibility with the currently available data. Hardin and Rosenberg, in a reply to Laudan, comment that successful theories 'are both comprehensive and robust, i.e., supported by convergent lines of independent argument'.[56] With this characterization in mind, some of Laudan's examples begin to seem controversial at the very least. Take his citation of the crystalline sphere theory of ancient and medieval astronomy. Hardin and Rosenberg's comment is:

That theory . . . started life as parasitic upon kinematic astronomy and never relinquished that role. As far as we can tell, it played no heuristic part in Eudoxus' solution to Plato's problem nor in the subsequent development of astronomy as an exact science.[57]

The point is that various theories which were compatible with the data were no more than that—they were not 'comprehensive and robust'. Success means more than fit.

McMullin similarly takes Laudan to task, denying that the realist will count very many of these theories as successful, or their 'success' as attributable to their theoretical assumptions. He writes:

The sort of theory on which the realist grounds his argument is one in which an increasingly finer specification of internal structure has been obtained over a long period, in which the theoretical entities function *essentially* in the argument and are not simply intuitive postulations of an 'underlying reality', and in which the original metaphor has proved continuously fertile and capable of increasingly further extension.[58]

McMullin's point is the entirely proper one that success for the realist implies more wide-ranging promise than Laudan seems to have allowed in drawing up his list, and that this success must be

[55] Ibid. 23.
[56] C. L. Hardin and A. Rosenberg, 'In Defense of Convergent Realism', *Philosophy of Science*, 49 (1982), 610.
[57] Ibid.
[58] 'A Case for Scientific Realism', 17.

closely linked to theoretical assumptions. This excludes, he maintains, not only the crystalline spheres of ancient astronomy, but also the universal deluge of catastrophist geology, theories of spontaneous generation, and fluid and ether theories. It may be, then, that with attention to the requisite historical details and with a proper concept of success, many of the theories on Laudan's list will be dispatched as *un*successful theories. But even if this does not account for all the theories, the realist would deny that Laudan's list is decisive in any case. For, as McMullin notes, continuity is more characteristic of science than is reversal and revolution, even if there are some successful theories which are not approximately true.

Laudan's challenge is that a theory must refer to be approximately true, and that some successful theories do not refer; hence there can be no inference from success to approximate truth. Allowing for the sake of argument that some of the theories cited by Laudan are genuinely successful, Hardin and Rosenberg suggest that the realist may yet question two further premises in Laudan's attack:

(1) that a theory must refer to be approximately true.
(2) that some successful theories do not refer.

In response to (1) the realist might adopt a view according to which a theory may be approximately true even though it does not refer. In response to (2) a notion of reference may be developed whereby reference can in fact be attributed to the range of previously successful theories which Laudan lists; for it may be that Laudan is sparing of attributing reference because he holds to a theory of reference which the realist repudiates, namely a descriptive theory. Hence his attributions do not so much show that none of these theories refers, but reflect an underlying prejudice that a theory which makes mistaken theoretical claims cannot do so. It may be that either of these two responses re-establishes the link between success and approximate truth on which the realist seems to rely.

Laudan remains unconvinced by the suggested replies and in an article entitled 'Realism Without the Real' contends that

the Hardin–Rosenberg approach buys what explanatory power it can claim at a very high cost; namely, by a more or less wholesale repudiation of much that the scientific realist holds dear. Indeed, it seems to me that the position they actually defend is so attenuated a form of 'realism'—if it be realism at all—that it is scarcely distinguishable from that of an instrumentalist.[59]

Hardin and Rosenberg, so Laudan alleges, 'seem to be cutting the ground out from under classical scientific realism'.

In relation to the first response to Laudan's list of successful theories—that we allow that theories which do not refer may none the less be approximately true—doubts would appear to be justified. If we say that a theory fails to refer it becomes difficult to say of the theory itself, as opposed to its predictions, that it is approximately true. It is hard to grasp the sense of granting approximate truth to a non-referring *theory*; a statement which fails to refer is more usually thought of either as lacking a truth value, or as being false.

Given that we agree with Laudan that the first response put into the mouth of the realist is indeed highly inappropriate, we may turn to the second. The point of this response is to call into question Laudan's claim that there are many successful theories which do not refer, by proposing the adoption of a 'charitable' theory of reference.

This is not the place to enter into a detailed discussion of the theory of meaning, but naturally enough the dispute with Laudan passes through this area. For any assertion that there are many successful but untrue (because non-referring) theories flows from judgements of referential success and failure which are in turn informed by a theory of reference. Or perhaps the relationship between theories of reference and our judgements concerning reference is more subtle than that last comment suggests. For in the first instance we do not turn to a theory of reference to help us decide whether two speakers are talking about the same thing. Rather it is the judgements about meaning we find natural and intuitively plausible which are primary, and which provide the measure of any conceptual account. That account should protect our intuitions. Should it prove impossible to do so, then there may be a case for revising our intuitions, but the realist will start out at least from what it seems right to say in relation to actual cases.

Take the case of the electron mentioned by Putnam and enlarged upon by Newton-Smith.[60] Thomson believed things about the

---

[59] L. Laudan, 'Realism without the Real', *Philosophy of Science*, 51 (1984), 156.

[60] Putnam, 'Explanation and Reference' in *Philosophical Papers*, ii. *Mind, Language and Reality* (Cambridge: Cambridge University Press, 1975), 197 f; and Newton-Smith, *Rationality of Science*, 158–64.

electron which Bohr denied, just as present-day scientists believe
things incompatible with the theoretical suppositions made by
Bohr. If one were to adopt (as does Kuhn) the holistic, or
descriptive conception of meaning which suggests that there is a
shift in meaning of 'electron' every time there is a change in the
theory, then it would follow that Bohr and Thomson were not even
talking about what we talk about when we mention electrons.
Indeed, it would seem that, since Bohr and Thomson were talking
about things which scientists now hold to be non-existent (since
nothing has the attributes they associated with electrons), they both
failed to refer. And if they failed to refer, then everything they said
about the electron is false or lacks a truth value. As Newton-Smith
observes 'it is at the very least offensive to be driven to the
conclusion that neither Thomson whom we think of as having
discovered the electron nor Bohr whom we think of as having made
important discoveries about the electron said anything true about
electrons.'[61]

Now this offensive conclusion is just what 'the traditional view',
as Putnam calls it, does drive us to. For one of the central
contentions of that view is, according to Putnam, 'That the
meaning of a term (in the sense of "intension") determines its
extension (in the sense that sameness of intension entails sameness
of extension).'[62] That is to say that the sense (expressed by a
description) fixes the reference (those things of which it is true).
Thus the sense of 'rabbit' might be, as in *The Concise Oxford
Dictionary*, 'burrowing . . . mammal . . . of hare family, brownish
grey in natural state, also black or white or pied in domestication'.
Its extension would be the set of things described by the sense.
And if that is the case then disagreement in sense produces
difference of reference and difference in meaning.

Putnam wants to turn this on its head. Rather than intension
enabling us to fix reference, it is extension which determines
intension. That is to say, the identity of terms in theories is
guaranteed not by their sense, but by their reference. Disagreement
in sense does not, then, produce difference of meaning, for there
may be 'trans-theoretical' terms. Thus common referential success

---

[61] *Rationality of Science*, 161.
[62] H. Putnam, 'The Meaning of "Meaning"', in *Philosophical Papers*, ii. *Mind,
Language and Reality*, 219.

may be achieved in spite of descriptive error and disagreement. But in that case the question arises as to how the reference of theoretical terms is fixed.

Putnam has elaborated a causal theory of reference, the details of which are controversial and need not concern us. The essential point is that reference may be fixed by, for example, an intention to pick out a natural kind or a physical magnitude. Take an example given by Hacking:

> In 1897 J. J. Thomson showed that cathode rays consist of what were then called 'ultratomic particles' bearing a minimal negative charge. These particles were for long called 'corpuscles' by Thomson, who rightly thought he had got hold of some ultimate stuff. He determined their mass. Meanwhile, Lorentz was elaborating a theory of a particle of minimum charge which he quickly called the electron. Around 1908 Millikan measured this charge. The theory of Lorentz and others was shown to tie in rather nicely with the experimental work.[63]

Each of the characters in the story had a different theory of the electron—some even used a different term. But we attribute to them shared reference for the reason that they intended to theorize about that which was responsible for the cathode ray phenomenon. There is here a causal description of the 'electron' which fixes the referent.[64] And, to cite Putnam again,

> once the reference is fixed, one can use the word to formulate any number of theories about that referent (and even to formulate theoretical definitions of the referent which may be correct or incorrect scientific characterizations of that referent), without the word's being in any sense a different word in different theories.[65]

Putnam's aim is to sketch a theory of meaning which will support the realist's insights that 'concepts which are not strictly true of anything may yet refer to something; and concepts in different theories may refer to the same thing.'[66] My brief account of his

---

[63] *Representing and Intervening*, 83.

[64] The word 'causal' is not being used in the manner intended by Putnam, whose view is that to establish common reference one needs to trace a present use back through a causal chain beginning with the original baptism or dubbing which introduced the term into the language. It is, instead, a causal theory in the sense that it maintains that the extension of a term may be specified as that which 'causes a certain phenomenon in certain ways'. See Newton-Smith, *Rationality of Science*, 174.

[65] 'Explanation and Reference', 202.

[66] Ibid. 197.

work does not, of course, constitute a fully adequate defence of it. It is, however, sufficient to suggest that the realist has an alternative to the descriptive theory of meaning with its counter-intuitive consequences. The causal theory avoids these consequences by loosening the connections between truth, approximate truth, and reference. It maintains that a theory may refer, be part of a progression of theories increasing in verisimilitude, and yet make numerous false theoretical claims. Indeed, theories near the beginning of the progression may well not be described as approximately true, for though referential success is a necessary condition for approximate truth, it is not itself sufficient.

Laudan does not discuss theories of reference directly, but he would not want to beg the question against the realist by assuming a theory of reference which the realist disputes. None the less it seems that his list of successful and non-referential theories does constitute a begging of the question. For if anything has emerged from the discussion of Putnam's work on reference, it is that determining the reference of a theory is not necessarily an easy task. It is not sufficient, as it would be if a descriptive account could be taken for granted, merely to latch on to what we would regard as theoretical errors to establish the referential failure of certain theories. Thus Hardin and Rosenberg are rightly critical of the confidence with which Laudan consigns certain theories to his list. For example, to them 'it seems not unreasonable . . . for realists to say that "ether" referred to the electromagnetic field'.[67]

This analysis may work for ether theories and for some of the others on Laudan's list. But to Laudan this seems to be a rearguard action of limited success. For what bothers him is that the realist's theory of reference severs reference from approximate truth in a way which is fatal to realism. Thus he writes:

By insisting that a theory may be highly successful and that its central concepts may successfully 'refer', even when there is nothing in the physical world relevantly like the entities postulated by the theory in question, Hardin and Rosenberg have conceded that the deep structural claims of a theory can be systematically wide of the mark, although the theory itself is highly successful at those levels where it can be tested. Which is to say that there is no co-variance between the empirical success of theories and the correctness of their deep-structural claims. But to grant that is just to

[67] 'In Defense of Convergent Realism', 613–14.

concede that 'inference to the best explanation', which is the realist's core inferential strategy, is badly flawed.[68]

Laudan is here threatening the realist with the following dilemma. Even if you have a theory of reference which is more charitable than the one I employ, it still must happen that there are theories which on your account are referring but not approximately true—this is in a way the whole point of the causal, non-descriptive theory of reference. However generous we are in attributing to theories referential success, not every theory which refers will be approximately true. Thus the Putnamesque theory of reference does not provide an escape from the charge that the inference from success to approximate truth is illicit, for it grants that there are successful but untrue (even if referring) theories in the history of science.

That is the nub of Laudan's objection to Hardin and Rosenberg's second way. As Laudan was eager to point out, the realist must have some response to the existence of previously successful theories. It has been argued that a more tolerant theory of reference is plausible, and that it opens up the possibility that a number of theories on Laudan's list may be held to refer, thereby meeting the minimum requirement for their being approximately true. But, so Laudan would argue, the theory of reference *does not* promise to allow us to explain the success of them all in terms of their approximate truth, a fact which renders the theory of reference largely irrelevant. The charitable principle of reference is required if realism is to avoid the conclusion that all successful past theories are plain false because non-referring, but it succeeds by separating referential success from theoretical accuracy. Hence it does not undermine the allegation that there are successful but untrue theories, and that, in consequence, success does not betoken verisimilitude.

Laudan's second objection to realism was based on the existence of successful but untrue because non-referring theories; hence it is illicit for the realist to infer from success to approximate truth, just because there are these untrue but successful theories. If theories can be successful and lack approximate truth, why should we accept that verisimilitude is our best explanation of success? The success of the theories on Laudan's list is not uncontroversial, but there are two answers which take it for granted. I have suggested

---

[68] 'Realism without the Real', 161.

that the first, which claims that theories may be approximately true
and yet not refer, is no answer at all. The second provides only the
start of a vindication of realism. It will require that the realist go
back to the history of science, to see whether the theories which are
successful are in fact theories so awry in their structural claims that
they cannot be counted approximately true, even if the realist
considers them to refer. If they are not so awry, their success may be
brought within the ambit of the realist's explanation.

There remains, however, a fourth reply to Laudan's objection
which takes the defence of realism further. Suppose that it proved
impossible to account for all the theories Laudan alleges to be
successful but not approximately true; imagine that suggesting that
the success of some was spurious, or that some do in fact refer, did
not exhaust the list Laudan gives. We are left with theories we have
to admit to be successful and in no sense approximately true, even if
we have accorded them the status of theories which refer. Laudan's
error lies in supposing that this eventuality would necessarily
embarrass the realist.

In fact the realist need have no cause for concern, for realism's
claim is not that success can only be explained by verisimilitude,
nor that it can always be so explained. The claim is that
approximate truth is, other things being equal, the most likely
explanation; the realist position is not undermined, therefore, by a
few examples where the explanation does not fit. It would be
reasonable, let us say, to infer of a particular case of lung cancer
that it is caused by smoking even though not all cases are so caused.
And it is reasonable to do so even if we will sometimes be led astray
by the inference. The fact that approximate truth is not a necessary
condition for success, just as smoking is not a necessary prelude to
lung cancer, does nothing to rebut the claim of the rational realist
that verisimilitude is explanatory of success. For though the realist
admits that a theory may be false and yet have true consequences,
the realist also claims that realism provides an account of the
systematic and sustained success of scientific theory. Even if we are
led astray by it in one or two cases, realism's central intuition
remains trustworthy, that the truth or verisimilitude of theories is
the best explanation of their success. For it is from a whole string of
true consequences, related one to another as the theories of science
are, that one infers from success to verisimilitude. It is this shared
success which the realist alleges cannot be coincidental.

The realist cause is not yet vindicated, however, for Laudan's final claim is that even if his case based on historical examples had not worked, still the realist argument involves a *petitio principii*. It begs the question, for the very point of the traditional anti-realist case is that it is wrong to affirm a theory merely on the ground that it has some true consequences. The whole dispute is about the very legitimacy of inference to the best explanation. Hence to appeal to the fact that realism is explanatory of success in a dispute about the validity of inference to the best explanation is circular; as Laudan puts it:

the new breed of realist . . . wants to argue that epistemic realism can reasonably be presumed to be true by virtue of the fact that it has true consequences. But this is a monumental case of begging the question. The non-realist refuses to admit that a *scientific* theory can be warrantedly judged to be true simply because it has some true consequences. Such non-realists are not likely to be impressed by the claim that a *philosophical* theory like realism can be warranted as true because it arguably has some true consequences. If non-realists are chary about first-order abductions . . . they are not likely to be impressed by second-order abductions.[69]

The same accusation is made by Arthur Fine, who notes that the anti-realism of Osiander, Poincaré, and Duhem was motivated by doubts about whether abductive inferences yield theories which are even approximately true. Thus 'one must not beg the question as to the significance of explanatory hypotheses by assuming that they carry truth as well as explanatory efficacy.'[70]

The realist wonders how it can be admitted that a theory is undoubtedly the best explanation, yet maintained that in itself this is no reason to believe it to be probably true or approximately true. The fact that a theory is successful is surely a reason (albeit fallible) for inferring this of it. So it seems best to reply to the charge that the realist's argument is circular or begs the question by insisting that the argument from the success of science is not supposed to warrant inferences to the best explanation for that was not its purpose. Rather its purpose is to show to those who accept this as a legitimate mode of argument that it warrants scientific realism. The form of the argument is not so much circular, then, as *ad hominem*. It is not a justification of abduction, but rather an application of it.

---

[69] 'Confutation', 45.
[70] 'Natural Ontological Attitude', 86.

Maybe Laudan's point is that, as an *ad-hominem* argument, it has limited force. But Boyd says this:

The rejection of abduction or inference to the best explanation would place quite remarkable strictures on intellectual inquiry. In particular, it is by no means clear that students of the sciences, whether philosophers or historians, would have any methodology left if abduction were abandoned. If the fact that a theory provides the best available explanation for some important phenomenon is not a justification for believing that the theory is at least approximately true, then it is hard to see how intellectual inquiry could proceed. Of course, the antirealist might accept abductive inferences whenever their conclusions did not postulate unobservables, while rejecting such inferences to 'theoretical' conclusions. In this case, however, the burden of proof would no longer lie exclusively on the realist's side: the antirealist must justify the proposed limitation on an otherwise legitimate principle of inductive inference.[71]

And he strengthens this argument with the observation:

unless—as is very unlikely—the apparent theory-dependence of inductive inference about observables is really only apparent, the empiricist who rejects abductive inferences regarding unobservables must hold that even the inductive inferences scientists make about observables are unjustified.[72]

If Laudan and Fine continue in their doubts about abduction they will end up with a severely atrophied view of the world, a view without many of the elements we take for granted. The strategy adopted by the realist is to point out the fundamental role played by inference to the best explanation in reasoning practices which should only be rejected with a full awareness of the dire epistemological consequences this rejection would entail. For these consequences are so dire as to be acceptable to none but the out-and-out sceptic. The realist, however, did not begin an argument with the sceptic, but with a more moderate form of anti-realism with its own distinctive set of worries. Answers to those worries have been offered. If the anti-realist now appears in the guise of sceptic, the realist will feel cheated, for the argument with the sceptic is a different sort of argument altogether, and not one which the rational realist has a particular responsibility to engage in.

---

[71] 'Current Status of Scientific Realism', 67.
[72] Ibid. 67–8.

PART II

The Rationality of Religious Belief

4.

Does Philosophy of Religion Rest on a Mistake?

1. On Justifying Religion

In Chapter 1 I drew attention to the frequently made claim that there is between scientific and religious belief a sharp dichotomy. It was also pointed out that a comparison between the two requires a philosophical analysis of both. In the last two chapters the outlines of an account of science have emerged as Kuhn's *Structure of Scientific Revolutions* has been criticized and a defence of rational realism has been offered. Rational realism is not without its problems, but I hope to have shown that it has a prima-facie plausibility which has given rise to its present popularity amongst philosophers of science. It is the view of science held by the rational realist which provides the point of comparison for religious belief.

In defending the rationality of science the rational realist brought to the fore a pattern of argument, inference to the best explanation, which is allegedly characteristic of science, and which may have an important role to play in the justification of theism. That it does have such a role has been maintained by, among others, Basil Mitchell, who thinks that theism is to be given what we might term an 'explanatory justification'. If Mitchell's claim is defensible, then a strong analogy between religious and scientific belief is established.

Mitchell asserts that the case for theism 'must be a cumulative one which is rational, but does not take the form of a strict proof or argument from probability'.[1] In this explanatory case the traditional proofs of the existence of God play a new role:

What has been taken to be a series of failures when treated as attempts at purely deductive or inductive argument could well be better understood as contributions to a cumulative case. On this view the theist is urging that traditional Christian theism makes better sense of all the evidence than does any alternative on offer, and the atheist is contesting the claim.[2]

[1] *Justification of Religious Belief*, 39.
[2] Ibid. 39–40.

Richard Swinburne's *The Existence of God* is a detailed defence of the assertion that theism 'makes better sense of all the evidence'. The evidence which is most important to the theist, according to Swinburne, is that which has provided the basis for many of the well-known a-posteriori arguments for the existence of God: 'the existence of the universe, its conformity to order, the existence of animals and men, men having great opportunities for co-operation in acquiring knowledge and moulding the universe, the pattern of history and the existence of some evidence of miracles, and finally the occurrence of religious experiences'.[3] Each one of these pieces of evidence, Swinburne argues, is more to be expected if God exists than if he does not. And taken together he thinks they render the existence of God more probable than not. He writes:

The phenomena which we have been considering are puzzling and strange. Theism does not make their occurrence very probable; but nothing else makes their occurrence in the least probable, and they cry out for explanation. *A priori*, theism is perhaps very unlikely, but it is far more likely than any rival supposition. Hence our phenomena are substantial evidence for the truth of theism.[4]

It is not my purpose to advance and test an explanatory justification (a cumulative case) for theism. Rather in subsequent chapters I shall discuss the religious appropriateness of the attempt to do so, and the conditions which such an explanatory justification would have to meet to be judged satisfactory. It will emerge that certain misgivings about the appropriateness and possibility of an explanatory justification rest on mistaken conceptions of the nature and practice of science.

The suggestion that religious belief might be given an explanatory justification supposes, in the first instance, that religious belief can properly be seen as functioning as an explanation. But this may be challenged at the outset. To defend religion as explanatory, it may be argued, is to ignore the true character of religious discourse which is expressive. Thus D. Z. Phillips, a leading exponent of this view, tells us: 'It is a grammatical confusion to think that this language is referential or descriptive. It is an expression of value. If one asks what it says, the answer is that it says itself.'[5] Hence the

---

[3] R. Swinburne, *The Existence of God* (Oxford: Clarendon Press, 1979), 277.
[4] Ibid. 290.
[5] D. Z. Phillips, *Religion without Explanation* (Oxford: Blackwell, 1976), 147.

project of justification, along with the tradition of philosophy of religion which contemplates such a project, mischaracterizes religious discourse.

Is religious discourse concerned primarily with describing and explaining (hence with making statements to be judged true or false), or is it primarily concerned with expressing and commending a particular attitude towards the world? For shorthand, we call these the 'intellectualist' and 'expressive' (or non-cognitive) accounts respectively. The project of justification presupposes that the first approach is correct, and the second in error.

In this chapter I shall begin by expounding an expressive understanding of religious discourse, giving examples of how particular 'beliefs' may be construed as expressions of value. I note that the expressive understanding is allegedly descriptive rather than normative, and consider two justifications for this claim suggested by Wittgenstein's writing. The justifications are based on what I term the arguments from the strong and weak principles of charity. The strong principle maintains that in interpreting belief one should maximize true beliefs; the weak principle, that one should maximize intelligible behaviour. Both principles allegedly justify the expressive understanding of religious discourse.

The first principle is itself indefensible, and so the case based on the second is alone to be taken seriously. It relies on an examination of the behaviour of believers, which behaviour, it is suggested, is rendered more intelligible by an expressive than by an intellectualist understanding. In particular, great stress is laid on the immunity of religious belief to evidence. Whilst it is admitted that the expressive interpretation has a prima-facie plausibility if the facts are as maintained, it is replied that aspects of behaviour associated with religious belief (above all loss of belief) are rendered unintelligible if it is maintained that religious belief is 'immune to evidence'. The great strength of the expressive thesis is in its making sense of the hesitation we feel in describing these beliefs as hypotheses; they seem to have a greater security in the believer's noetic structures than that description implies. But that fact can be coped with, so I shall argue in Chapter 5, not by denying that these beliefs have anything to do with evidence, but rather by showing how it is that beliefs resting on an explanatory justification may legitimately possess a certain tenacity in the face of conflicting evidence.

2. *The Expressive Theory and its Defence*

It was a leading exponent of intellectualism at the turn of the century, Sir James Frazer, who provoked Wittgenstein's *Remarks on Frazer's* Golden Bough. Wittgenstein's contempt for *The Golden Bough* was intense:

What narrowness of spiritual life we find in Frazer! As a result: how impossible for him to understand a different way of life from the English one of his time! Frazer cannot imagine a priest who is not basically an English parson of our times with all his stupidity and feebleness.[6]

And again,

Frazer is much more savage than most of his savages, for these savages will not be so far from any understanding of spiritual matters as an Englishman of the twentieth century. His explanations of the primitive observances are much cruder than the sense of the observances themselves.[7]

Wittgenstein's objection is to Frazer's intellectualism, his account of religion and magic which 'makes these notions appear as *mistakes*', that is, as misguided attempts at explanation.[8] In contrast, Wittgenstein thinks that 'what makes the character of ritual action is not any view or opinion, either right or wrong', and that understanding of such rituals comes not from historical conjectures and the like, but by 'perspicuous presentation'.[9] As he puts it:

I think one reason why the attempt to find an explanation is wrong is that we have only to put together in the right way what we *know*, without adding anything, and the satisfaction we are trying to get from the explanation comes of itself.[10]

The sense of these observances will be found to lie in the nature of human life and emotions, rather than in speculative, explanatory theories.

In *Lectures and Conversations on Aesthetics, Psychology and Religious Belief* Wittgenstein approaches religious belief (Christianity in particular) in the same spirit, rejecting the idea that a religious

---

[6] L. Wittgenstein, *Remarks on Frazer's* Golden Bough, ed. R. Rhees, trans. A. C. Miles (Retford, Notts.: Brynmill Press, 1979), 5. For this discussion there is no need to question whether Wittgenstein correctly represents Frazer's position.
[7] Ibid. 8.   [8] Ibid. 1.
[9] Ibid. 7 and 9.   [10] Ibid. 2.

way of life rests on a metaphysical foundation. Discussing the belief that there will be a last judgement, he writes:

Here believing obviously plays much more this role: suppose we said that a certain picture might play the role of constantly admonishing me, or I always think of it. Here, an enormous difference would be between those people for whom the picture is constantly in the foreground, and the others who just didn't use it at all.[11]

And later, throwing light on this, he writes that 'In a religious discourse we use such expressions as: "I believe that so and so will happen," and use them differently to the way in which we use them in science.'[12]

D. Z. Phillips has applied Wittgenstein's hints systematically to develop an expressive understanding of Christian discourse. Taking up Wittgenstein's words in *Lectures and Conversations*, he notes that religious pictures have a distinctive role which sets them apart from other pictures:

if someone used the pictures of the plants as proof of the reality of plants someone might say, with justification, 'I shan't be convinced if you can only show me these pictures. I shall only be convinced when I see the plants.' If, on the other hand, having heard of people praising the Creator of heaven and earth, glorifying the Father of us all, feeling answerable to the One who sees all, someone were to say, 'But these are only religious perspectives, show me what they refer to', this would be a misunderstanding of the grammar of such perspectives. The pictures of the plants refer to their objects, namely, the plants. The religious pictures give one a language in which it is possible to think about human life in a certain way. . . . [they] provide the logical space within which such thoughts can have a place. When these thoughts are found in worship, the praising and the glorifying does not refer to some object called God. Rather, the expression of such praise and glory is what we call the worship of God. . . . religious expressions of praise, glory, etc. are not referring expressions. These activities are expressive in character, and what they express is called the worship of God.[13]

Since the demand 'show me what they refer to' is out of place, Phillips is led to the conclusion that philosophy of religion as it is

[11] L. Wittgenstein, *Lectures and Conversations on Aesthetics, Psychology and Religious Belief*, ed. C. Barrett (Oxford: Blackwell, 1966), 56.

[12] Ibid. 57.

[13] *Religion without Explanation*, 148–150. See Wittgenstein, *Lectures and Conversations*, 57.

traditionally practised (with its various searches for, and denials of
the possibility of, a proof of the existence of God), rests on a
mistake; for 'where religious belief is concerned, to speak of proof
and explanation is to betray a misunderstanding of what is being
investigated.'[14]

In further clarification of this position, Phillips has provided an
explication of various Christian beliefs as they appear to the
expressive theory, in particular beliefs regarding the nature of
prayer and the afterlife.[15] Of prayer Phillips gives the following
account:

When deep religious believers pray *for* something, they are not so much
asking God to bring this about, but in a way telling Him of the strength of
their desires. They realize that things may not go as they wish, but they are
asking to be able to go on living whatever happens. In prayers of confession
and in prayers of petition, the believer is trying to find a meaning and a
hope that will deliver him from the elements in his life which threaten to
destroy it: in the first case, his guilt, and in the second case, his desires.[16]

It should be borne in mind just how minimal is the notion of
'asking' here, notwithstanding the traditional phrasing.[17] For
though the person who prays is asking 'that his desire will not
destroy the Spirit of God within him', for Phillips there is no one
addressed here who might respond in some way, for there is
nothing to which 'God' refers. Such prayer is not too dissimilar
from hoping, and hoping which is just plain hoping, without any
foundation for confidence. It presupposes no factual beliefs which
may turn out false, but is merely expressive of a reflective or
contemplative attitude towards the world.

In *Death and Immortality* Phillips argues that immortality need
not be understood as a prediction of life after death, but can be
construed as a vantage point from which one sees one's life:
'Eternity is not an extension of this present life, but a mode of
judging it. Eternity is not *more* life, but this life seen under certain
moral and religious modes of thought.'[18] And in a chapter entitled

[14] *Religion without Explanation*, 43.
[15] D. Z. Phillips, *The Concept of Prayer* (London: Routledge & Kegan Paul,
1965) and *Death and Immortality* (London: Macmillan, 1970).
[16] *Concept of Prayer*, 121.
[17] This is stressed by W. D. Hudson, 'Some Remarks on Wittgenstein's Account
of Religious Belief', in *Talk of God*, ed. G. N. A. Vesey (Royal Institute of
Philosophy Lectures, 2; London: Macmillan, 1969), 46–8
[18] *Death and Immortality*, 49.

'Perspectives on the Dead' Phillips considers the claim that beliefs about the dead must rest on some 'fantasy', such as that the dead are now alive but in a way different from how the rest of us are alive.[19] Against such beliefs he prefers the views of Peter Winch and Simone Weil, for they offer ways of thinking about life after death which do 'not presuppose . . . a belief that those we thought were dead are in fact still living'. Their beliefs are of 'the dead as being present in the form of absence'.[20]

Here, then, we have an expressivist account of particular religious beliefs, an account which we might be tempted to take as normative, and thus no different in kind from the proposals for a new sort of religious belief offered by, for example, Braithwaite, Hare, and Sutherland.[21] At times Phillips seems to be encouraging us to accept his account as normative, for he points out that religious realism is philosophically and spiritually indefensible, which is not at all the same as saying that religious realism is untrue to religious belief. If his point is that the expressive account is to be preferred to a realist one given the philosophical and spiritual drawbacks of the alternative, his own argument is not so much a challenge to the descriptive accuracy or relevance of the traditional conception, but a new interpretation of religion which is offered when the old one has been rejected.

But the expressive account is not to be understood in that way. Wittgenstein does not couch his observations in the form of advice; that is to say, they have the form not of prescriptions, but of descriptions. He writes of religious beliefs and practices that 'what we have here is *not* an error', implying that Frazer's interpretation which makes them so is mistaken.[22] So too, Phillips claims that his is a descriptive, rather than a normative account.[23] Religious language is not even putatively fact-stating, and hence to talk of

---

[19] *Religion without Explanation*, 122–38.

[20] Ibid. 129.

[21] R. B. Braithwaite, 'An Empiricist's View of the Nature of Religious Belief', in *The Philosophy of Religion*, ed. B. G. Mitchell (Oxford: Oxford University Press, 1971), 72–91; R. M. Hare, 'Theology and Falsification', in *The Philosophy of Religion*, ed. Mitchell, 15–18; and S. Sutherland, *God, Jesus and Belief* (Oxford: Blackwell, 1984).

[22] *Remarks on Frazer*, 2.

[23] Phillips assumes that Hume put paid to any attempt to defend philosophical theism, and this is an assumption which plays a part in his case. But that is not *why* Phillips adopts his position, so he tells us, his purpose being rather to explicate the true nature of religious belief. See *Religion Without Explanation*, p. ix.

justification and proof is to import what is appropriate in one context to another where it is not, betraying confusion and misunderstanding.[24] Construing religious language in this way, Phillips is able to turn his critics' criticism back on themselves:

> Is it reductionism to say that what is meant by the reality of God is to be found in certain pictures which say themselves? If we mean by reductionism an attempt to reduce the significance of religious belief to something other than it is, then reductionism consists in the attempt, however sophisticated, to say that religious pictures must refer to some object; that they must describe matters of fact. That is the real reductionism which distorts the character of religious belief.[25]

Such, then, are the claims of the expressive theory. How are they to be defended? There is a tendency to treat with scorn and as wholly implausible the notion that the expressive account is descriptive, as if this were little more than a rhetorical device. In exemplifying this attitude, there is common ground between Swinburne and J. L. Mackie. Thus Swinburne comments:

> Phillips's account of prayer is . . . best regarded as a suggested reinterpretation of prayer for those who feel that, even if there is no transcendent God, there is still a point in pursuing an activity, which is not prayer, but which, in the words and gestures used has some outward similarity to prayer.[26]

And Mackie notes of Phillips's explication of Christian beliefs concerning immortality:

> Stripped of some romantic phrases, this means simply that the dead survive only in the thoughts of those who remember and miss them, but that there is nothing absurd in remembering and missing one's dead friends, though coming to terms with the void created by their deaths. . . . Phillips is trying to take over, in the name of religion, not only the factual beliefs but also the moral attitudes and values of atheism.[27]

---

[24] According to Phillips the ranks of the confused include Mascall, Hick, Hepburn, Crombie and Mitchell: see D. Z. Phillips, *Faith and Philosophical Enquiry* (London: Routledge & Kegan Paul, 1970), 1, 5, 13, 72, 129; and *Concept of Prayer*, 88–94.

[25] *Religion without Explanation*, 150. The notion of pictures 'saying themselves' is a recurrent theme in Phillips's work; its meaning is explained in the comment: 'The picture is not a picturesque way of saying something else. It says what it says, and when the picture dies, something dies with it, and there can be no substitute for that which dies with the picture.' *Faith and Philosophical Enquiry*, 119.

[26] R. Swinburne, *Faith and Reason* (Oxford: Clarendon Press, 1981), 141.

[27] J. L. Mackie, *The Miracle of Theism* (Oxford: Clarendon Press, 1982), 229.

Swinburne and Mackie join in the common reaction that there is something fishy about the claim of expressivism to be descriptive, for we are inclined to take for granted that religious beliefs usually involve the making of factual claims. Yet such a reaction is unsatisfactory. For what are we then to make of the assertion that the expressive account is descriptive? Either that Wittgenstein and Phillips are in bad faith in employing a rhetorical trick or, more generously, that they are deluded. The undesirability of imputing either fault to either philosopher leads me to consider whether there are arguments which they might deploy in support of their position. Such reflection reveals two possible lines of defence suggested by their work. (And it also has the merit that it provides an explanation of why—when it is initially surprising—they seem unconcerned to distinguish magic, witchcraft, and religion from one another, but rather treat them as if they are in important respects alike.)

The first argument in defence of the expressive interpretation of religious discourse I label 'the argument from the strong principle of charity'. It takes seriously the uncompromising nature of Wittgenstein's assertions with respect to the character not of *some* religion and magic, but of religion and magic in general, of which he writes 'what we have here is *not* an error', for, as he comments further on, 'a religious symbol does not rest on any *opinion*.'[28] 'It is very queer', he says, 'that all these practices are finally presented [by Frazer], so to speak, as stupid actions. But it never does become plausible that people do all this out of sheer stupidity.'[29] These strong claims for the expressive account rely on Wittgenstein's argument that religion and magic would be mistakes if understood as explanatory, but that it is never plausible to attribute to believers such sustained and systematic errors: 'You might say: "For a blunder, that's too big." '[30]

Bearing in mind the form of the argument from the strong principle of charity, we see at once the significance for the expressive case of magic, witchcraft and so on. For the proposal that religious beliefs are really expressive is supported by the contention that religious realism is implausible—too big to be a blunder. Apart from disputing the claim that religious realism is implausible, an obvious reply would be that it is not in the least

[28] *Remarks on Frazer*, 2 and 3.
[29] Ibid. 1.
[30] *Lectures and Conversations*, 61–2.

surprising that people should commit such blunders. We have only to look at the history of magic and religion to find evidence of great credulity. Why should credulity be any less prevalent now? To counter this reply, the argument from the strong principle necessarily takes on a certain generality. It has to maintain that imputing false beliefs to believers is *always* a misguided explanatory tactic, whether the beliefs concern magic, witchcraft, or theism.

Let us suppose that Wittgenstein is relying on a strong principle of charity in the interpretation of belief, namely that it is not explanatory to impute false beliefs to others. Such a principle would resemble the strong principle of charity which some have claimed to find in Davidson's work on translation and interpretation. The strong principle suggests that in interpreting the beliefs of others we should maximize true beliefs; that we should make p's beliefs as much like ours as possible. But prompted by criticism from Lewis, Davidson has proposed instead the maximizing of intelligible behaviour as the constraining principle on interpretation. And this is clearly right: sometimes it will be more plausible that p believes falsehoods, given the circumstances, than that p shares our beliefs. Otherwise it would be sufficient to show that beliefs are mistaken in order to demonstrate that they could not be held.[31]

The strong principle of charity will not do. The mere fact that beliefs will be in error if interpreted in a particular way is not sufficient to justify an alternative interpretation. It may, however, give an alternative interpretation a prima-facie plausibility. And this modification suggests a second weak principle of charity which might help Wittgenstein and Phillips: that one should maximize intelligible behaviour. Since in construing specific beliefs Wittgenstein and Phillips both draw attention to aspects of behaviour associated with spiritual forms of life, it cannot be that they mean to assert the strong principle of charity, the implications of Wittgenstein's generalizations notwithstanding.

What I shall label 'the argument from the weak principle of charity' claims that in the interpretation of belief one cannot ignore the evidence of behaviour. It is a principle of charity in that it

[31] See David Lewis, 'Radical Interpretation', *Synthèse*, 23 (1974), 331–44; and Donald Davidson, 'Replies to David Lewis and W. V. Quine', *Synthèse*, 23 (1974), 345–9.

assumes a coherence between belief and behaviour, and it suggests a way in which the expressivist account might be defended. For this different, weaker principle of charity urges that it is not just the form of what people say which counts in determining what they mean, but also what they do. As Wittgenstein writes in *On Certainty*, 'Our talk gets its meaning from the rest of our proceedings.'[32] Now it may be argued in relation to religious assertions that 'the rest of the proceedings' of religious people demonstrate that religious belief is best understood expressively.

In his *Lectures and Conversations* Wittgenstein imagines someone going to a blackboard and saying '2 and 21 is 13'. He writes:

There are cases where I'd say he's mad, or he's making fun. Then there might be cases where I look for an entirely different interpretation altogether. In order to see what the explanation is I should have to see the sum, to see in what way it is done, what he makes follow from it, what are the different circumstances under which he does it, etc.[33]

The story is designed to persuade us of a problem in the interpretation of religious belief, namely that in seeing what is meant by a sentence it is necessary to see the language game of which it is a part. Talk of God's existence may have superficial similarities with talk of the existence of giant pandas, but the depth grammar may be different. If it is, attention to the verbal form of a sentence will not be sufficient to settle its meaning. One has to determine the context of the sentence and what does and does not follow from it, if one is to understand its meaning. Otherwise one will be uncharitable in interpreting the relationship between beliefs and behaviour.

Mackie comments on the expressive understanding that: 'it would be rather surprising if sentences which on the surface have reasonably straightforward factual, descriptive meanings had, in a certain kind of use, *only* . . . non-descriptive meanings.'[34] But if the weak principle of charity is legitimate, and hence not only form but also context is significant in interpreting language, it should not be surprising. Thus the argument from the weak principle of charity makes sense of the expressive claim as more than a rhetorical trick. Against the intellectualist who holds that in general religious and

---

32 *On Certainty*, para. 239.
33 *Lectures and Conversations*, 62.
34 *Miracle of Theism*, 220.

magical beliefs should be seen as attempts at explanation, Wittgenstein and Phillips argue that it is far better to see these beliefs as expressive of what goes deep in people's lives. For look at what believers say *and* do. It then becomes apparent, if you make reference to the entire context, that these beliefs are not factual at all, for that is not how they are treated in 'the other proceedings'. Thus the expressive account (which rests not on the false principle of maximizing true belief, but on the legitimate principle of maximizing intelligible behaviour) is to be preferred to the intellectualist one.

An immediate reply which the intellectualist might make would be to say that part of the context of religious beliefs is what people themselves say about them—and do they not, on the whole, understand them as the intellectualist supposes? Of course part of the behaviour of which we have to make sense is what people say about their beliefs. Even so, that would not be an infallible guide to what their beliefs really are, since it is only part of the evidence. The other part of the evidence is how they treat their beliefs in other ways. Just as we would not take someone's describing himself or herself as 'courageous' as decisive if nothing else supported the claim, neither should we take someone's understanding of his or her beliefs as decisive if it is incompatible with other evidence. Hence Phillips's notion of people having mistaken religious beliefs is not to be ruled out as absurd, for it is not impossible that one may offer incorrect rationalizations of what one says and does.

Since this objection is not sufficient to dispatch the case for expressivism based on the weak principle of charity, we are bound to look at what Wittgenstein and Phillips find in the evidence to justify the expressive understanding of religion, magic, and witchcraft. To see how the case might be made in detail, Wittgenstein's *Remarks on Frazer's* Golden Bough has to be examined.

3. Intellectualism: The Theory of 'Primeval Stupidity'[35]

Wittgenstein ridicules the assumption of Frazer's *Golden Bough* that religion and magic are just bad science: 'It is very queer that all

[35] The description of the intellectualist understanding of religion and magic as the theory of 'primeval stupidity' is Cassirer's, cited by Phillips, *Religion without Explanation*, 32.

these practices are finally presented, so to speak, as stupid actions. But it never does become plausible that people do all this out of sheer stupidity.'[36] Intellectualism makes religion and magic matters of 'sheer stupidity' by interpreting them as involving beliefs which are unjustified. But against interpreting them in this way, Wittgenstein provides a battery of arguments which I separate under the following headings:

1. *The immunity to criticism argument* which speaks for itself:

It may happen, as it often does today, that someone will give up a practice when he has seen that something on which it depended is an error. But this happens only in cases where you can make a man change his way of doing things simply by calling his attention to the error. This is not how it is in connexion with the religious practices of a people; and what we have here is *not* an error.[37]

2. *The technical competence argument.* Wittgenstein draws our attention to a fact which it is easy to overlook, namely the ability of 'primitive' people in numerous practical arts:

The same savage, who apparently in order to kill his enemy, sticks his knife through a picture of him, really does build his hut of wood and cuts his arrow with skill and not in effigy.[38]

The purport of the argument is this: that someone who is aware of the need to cope with the world by using practical skills cannot be resorting to magic or religion to achieve the same technical ends, otherwise magic or religion would be employed in both cases. The same goes even more for people of the present day in a society with greater technological achievements. When prayers are directed to the Virgin Mary to save a relative from death, surely this is not thought of as an alternative to administering medicine?

3. *The due season argument* follows on from the last point and suggests that if rituals and so on are understood as supplements to other causal activities, then surely people would turn to them *whenever* they were faced with a difficulty. 'How useful it would be', comments Phillips, 'if rain could be obtained out of the rainy season.'[39] But rituals are not used in this way, so Wittgenstein alleges:

[36] *Remarks on Frazer*, 1.
[37] Ibid. 2.
[38] Ibid. 4.
[39] *Religion without Explanation*, 34.

I read, amongst many similar examples, of a rain-king in Africa to whom the people appeal for rain *when the rainy season comes*. But surely this means that they do not actually think that he can make rain, otherwise they would do it in the dry periods in which the land is 'a parched and arid desert'. For if we do assume that it was stupidity that once led the people to institute this office of Rain-King, still they obviously knew from experience that the rains begin in March, and it would have been the Rain-King's duty to perform in other periods of the year. Or again: towards morning, when the sun is about to rise, people celebrate rites of the coming day, but not at night, for then they simply burn lamps.[40]

One might make the same point about Christian prayers. Christians pray for a good harvest, but only when the crops are being planted, or when they are about to be gathered. They never pray that the fields will continue to bear fruit throughout the winter months. And to Wittgenstein this would suggest that they cannot be thinking that their prayers or ceremonies are somehow instrumental in obtaining rain, harvests, and the like, but rather that they are expressive of hopes and fears.

4. *The obviousness of error argument*: if religion and magic rest on some sort of error, as the intellectualist supposes, how is it that it is not noticed? Phillips refers to Tylor as having various suggestions to make (such as that the magician relies on skills other than magic, that the people in question are of low intelligence, and the like),[41] whilst Frazer points out that sooner or later a ceremony to bring rain just has to be successful. 'But then', Wittgenstein objects, 'it is queer that people do not notice sooner that it does rain sooner or later anyway.'[42]

Reviewing Wittgenstein's case against Frazer, Phillips asks how ritualistic and religious activities are to be understood if not as mistaken hypotheses. The answer to intellectualism is that 'The rituals can be seen as a form of language, a symbolism in their own right; a language and a symbolism which are expressive in character.'[43]

In a discussion of the problem of interpreting the beliefs of the Azande concerning witchcraft, Peter Winch makes a case against intellectualism parallel to Wittgenstein's.[44] Here the objects of

---

[40] *Remarks on Frazer*, 12.
[41] *Religion without Explanation*, 29.
[42] *Remarks on Frazer*, 2.
[43] *Religion without Explanation*, 35.

criticism are Evans-Pritchard and MacIntyre, neo-intellectualists who show their colours by suggesting that Zande magic is false. In attacking this instance of the mistaken hypothesis view, it seems to me that Winch makes two main points in favour of an expressive interpretation.

The first point—let us term it the *magic as prescriptive argument*—rests on the claim that:

The spirit in which oracles are consulted is very unlike that in which a scientist makes experiments. Oracular revelations are not treated as hypotheses and, since their sense derives from the way they are treated in their context, they therefore *are not* hypotheses. They are not a matter of intellectual interest but the main way in which Azande decide how they should act.[45]

The pronouncement of the oracle, so we are told, is a guide for conduct like the commandment 'love your father and mother'. It cannot turn out false.

The second point Winch makes I shall call the *indifference to contradiction argument*; it is similar to the first of Wittgenstein's observations, and stems from Winch's suggestion that in getting a 'foothold' for an interpretation of Zande beliefs we should take note of their indifference to what we would count as contradictions in their system. According to Evans-Pritchard the Azande hold that witchcraft is an inherited trait, detectable by *post mortem* examination. Now since they are a small and closely linked tribe, it would seem to us that a few positive instances of witchcraft would show all of them to be witches (since they are all related one to another), and a few negative instances, that none are witches. So why is this contradiction not a source of concern to them? Evans-Pritchard accounts for it by saying that 'they have no theoretical interest in the subject and those situations in which they express their belief in witchcraft do not force the problem upon them.'[46]

Now Winch finds it significant that, in Evans-Pritchard's phrase, 'they have no theoretical interest in the subject'.

---

[44] P. Winch, 'Understanding a Primitive Society', in *Ethics and Action* (London: Routledge & Kegan Paul, 1972), 8–49.

[45] Ibid. 20.

[46] Cited by Winch, ibid. 24. Some would claim that the same could be said of the belief of the average Christian in the Trinity.

This suggests strongly that the context from which the suggestion about the contradiction is made, the context of our own scientific culture, is not on the same level as the context in which the beliefs about witchcraft operate. Zande notions of witchcraft do not constitute a theoretical system in terms of which Azande try to gain a quasi-scientific understanding of the world. This in its turn suggests that it is the European, obsessed with pressing Zande thought where it would not naturally go—to a contradiction—who is guilty of misunderstanding, not the Zande. The European is in fact committing a category-mistake.[47]

Winch's conclusion is one shared with Phillips; that is, that the intellectualists ignore the fact that: 'Magical rites constitute a form of expression in which . . . possibilities and dangers may be contemplated and reflected on—and perhaps also thereby transformed and deepened.'[48]

4. Assessing the Argument for the Expressive Theory

The arguments as we found them in Wittgenstein and Winch may be listed as follows:

(i)   the immunity to criticism argument;
(ii) the technical competence argument;
(iii) the due season argument;
(iv) the obviousness of error argument;
(v)   the prescriptivity of magic argument;
(vi) the indifference to contradiction argument.

Granting for the moment the empirical assumptions implicit in these arguments, the question I want to ask is: 'How great a weight do they give to an expressive theory?' That is to say, other things being equal and the facts being as the advocates of the expressive theory suppose, do we do well to hold that magic and religion are non-cognitive? I shall argue that the points made by Wittgenstein and Winch, whilst not conclusive evidence for the expressive theory, taken together are persuasive. However, I shall then go on to question some of the implicit empirical assumptions in the argument as it relates in particular to Zande magic and Christian theism.

[47] Winch, 'Understanding a Primitive Society', 26.
[48] Ibid. 41.

Arguments (i) and (vi) can be treated as one. Assuming that believers in magic and religion are unconcerned when confronted by what the scientific outlook sees as weaknesses in their beliefs or by what looks like a contradiction, does this prove that they are not understood as quasi-scientific beliefs? Indifference in this matter may well be suggestive, but it is by no means a proof. Many people believe things at the same time as the evidence speaks loudly against them: for example, that if only people want to work, they can find a job. This last belief is not expressive merely because absurd.

The argument from technical competence, (ii), proposes that people who are accustomed to meet the demands of existence by technical accomplishments cannot think of magic as aimed at the same end, otherwise they would have no need of one or the other. As with the last argument, more needs to be said for this conclusion to follow, for religion and magic often concern those areas of people's lives over which there is no technical control, such as death, major illness, and the elements. The total inappropriateness or futility of technical endeavour in dealing with these problems directly may argue that a supplement to technical skill is just what is required. At this frontier the utility of human technology ends, and there begins the appeal to spirits, saints, and gods.

The due season argument has as initial appeal, for it is indeed striking, in Wittgenstein's sardonic phrase, that at night 'they simply burn lamps'. Just how many magical and religious ceremonies this point applies to is quite another question. Presumably in relation to illness the 'request' for a cure does not always presage recovery as the 'request' for rain always precedes the rainy season. Yet even if it is true that people perform ceremonies of 'request' only when they have reason to expect what they supposedly ask for, this does not of itself establish that the expressive interpretation is the true one. Imagine that I have a very rich uncle whose custom it is to give presents to his nephews and nieces solely on the occasion of their birthdays. Suppose too that he enjoys the power which money brings, and even on birthdays will give the present only if he is asked for it. The mere fact that I refrain from asking for a present on all but one day of the year does not show that I doubt *either* my uncle's ability to withhold the present, *or* that my asking for it is instrumental in obtaining it. A similar thing may be true in the case of aptly timed ceremonies for rain or harvest. The Christian might say that God will provide a plentiful harvest at the ordained and fitting time, but only when prayers are made. Hence though prayers for harvest should not be addressed to God in December, let us say, it is still the case that the request is, in the spring, instrumental.

The argument from the obviousness of error holds that if religion and magic were just a tissue of mistakes as Frazer thinks, then it is highly unlikely that they would have survived so long. This point suffers from the same weakness as (i) and (vi), that people do seem to have a tendency to believe even where the evidence is slim, and especially where the error is not patent. And further, like (i) and (vi) it rests on the unsubstantiated assumption that although the 'error' has become manifest, belief has not declined.

Argument (v) says that magical and religious beliefs are treated not as hypotheses, but as guides for life. But this is clearly a false dichotomy: the English avidly watch weather forecasts and may plan the very details of their day-to-day lives accordingly, but this does not show that weather forecasts are expressive and not hypotheses.[49]

5. Expressive or Intellectualist?

The case for the expressive theory goes like this: given the way in which believers behave towards their beliefs, the expressive interpretation is to be preferred, since it maximizes intelligible behaviour. If we accept an intellectualist account we have to explain why the beliefs are not treated like ordinary factual beliefs, otherwise this behaviour remains unintelligible.

In discussing the arguments one by one, I have noted that none is conclusive. That is to say, that the evidence cited is compatible with either the intellectualist or expressive account. But that is hardly to the point, it might be said. In interpreting belief we have to weigh competing explanations, and surely, taken together, the points in favour of the expressive account make more sense of the whole and constitute a cumulative case in its favour. Finding 'belief' associated with such behaviour, would one not be inclined to accept an expressive interpretation?

It would seem that that has to be conceded. Given the case Wittgenstein and Winch make against intellectualism (always

[49] Charles Taylor argues against the same false dichotomy in 'Rationality', in *Rationality and Relativism*, ed. Hollis and Lukes, 87–105.

assuming the facts are as they say they are), the charge that Frazer lacked a spiritual sense has some substance. The possibility that religious beliefs, practices, and rituals are expressive is one which Frazer would at least have done well to consider and in certain cases accept. Take the ceremony of adoption which Frazer describes:

> The same principle of make-believe, so dear to children, has led other peoples to employ a simulation of birth as a form of adoption. . . . A woman will take a boy whom she intends to adopt and push or pull him through her clothes; ever afterwards he is regarded as her very son, and inherits the whole property of his adoptive parents.[50]

Wittgenstein remarks that 'it is crazy to think that there is an *error* in this and that she believes that she has borne the child.'[51] He may not mean to imply that Frazer believed that, but clearly he wishes to insist that a ceremony such as this should have impressed upon Frazer the need for an expressive interpretation of certain of the practices of magic and religion.

Frazer had, however, only to consider aspects of our own culture to find expressive rituals, ones which far from involving false hypotheses, proclaim 'what goes deep in our lives'. An example given by Phillips is that of the practice of accompanying the dead to their graves, which clearly does not rest on some theory or hypothesis, but is a mark of our love and respect. It is not done for an end, but expresses an attitude to life and death.[52] Wittgenstein draws attention to similar rituals:

> Burning in effigy. Kissing the picture of a loved one. This is obviously *not* based on a belief that it will have a definite effect on the object which the picture represents. It aims at some satisfaction and it achieves it. Or rather, it does not *aim* at anything; we act in this way and then feel satisfied.[53]

If further evidence of the resourcefulness of the expressive account is needed, it is shown in relation to a custom which leaves MacIntyre saying that it is 'thoroughly incoherent'.[54] Among aborigines is found the practice of carrying about a stick or a stone and treating it as if it is, or embodies, the soul of the individual who carries it. If the stone is lost then the individual anoints himself as

[50] Cited in *Remarks on Frazer*, 4.
[51] Ibid.
[52] *Religion without Explanation*, 36.
[53] *Remarks on Frazer*, 4.
[54] Cited by Winch, 'Understanding a Primitive Society', 45.

the dead are anointed. Winch finds the conclusion of incoherence lame, for:

Consider that a lover in our society may carry about a picture or lock of hair of the beloved; that this may symbolize for him his relation to the beloved . . . Suppose that when the lover loses the locket he feels guilty and asks his beloved for her forgiveness: there might be a parallel here to the aboriginal's practice of anointing himself when he 'loses his soul'. And is there necessarily anything irrational about either of these practices? Why should the lover not regard his carelessness in losing the locket as a sort of betrayal of the beloved? . . . The aborigine is clearly expressing a concern with his life as a whole in this practice; the anointing shows the close connection between such a concern and contemplation of death.[55]

The conclusion we should draw is that if the principle of interpretation proposed by intellectualism is understood as implying that *all* magical, ritualistic, and religious practices and behaviour should be seen as relying on explanatory hypotheses, then it is a bad principle of interpretation. In looking at our own lives, as well as at primitive practices, it seems that an expressive account of certain rituals will be essential if we are to render aspects of behaviour intelligible. Frazer, as read by Wittgenstein, was pushing religion and magic into preconceived patterns, rather than those suggested by the whole context of those practices.

A preliminary conclusion is that an exclusively intellectualist approach to the understanding of religion and magic is misguided. But that is not the issue which is of most concern to us, which is rather the validity of an exclusively expressivist understanding. If the evidence is always as it is said to be, the expressive case is prima-facie plausible. But in any particular case what we need to do is to consider the whole array of evidence, taking Wittgenstein's arguments as a warning against too ready and uncritical acceptance of an intellectualist interpretation. Whether or not we should accept the prima-facie argument depends on the empirical accuracy of the evidential assumptions implicit in Wittgenstein and Phillips's case. What I want to do is to take two examples, that of Zande witchcraft and Christian belief, and see whether they are best understood as expressive.

[55] Winch, 'Understanding a Primitive Society', 45–6.

6. *The Expressive Theory and the Facts*

The expressive interpretation of magic and religion rests on the claim that, given the behaviour of believers, its own interpretation of their beliefs is to be preferred to an intellectualist one, for otherwise the behaviour becomes unintelligible and we are committed to belief in widespread error. It is not that people cannot be stupid, but that by their behaviour they reveal that they are not. I have argued that these points do not constitute a proof, but strongly support an expressive account. An important assumption of the case has, however, so far gone unchallenged; that is, that the character of the behaviour of devotees of magic and religion is as Wittgenstein and Phillips say it is. But ought we not to question the claim that, for example, believers really take no interest in contradictions and counter-evidence?

Returning to Winch's essay 'Understanding a Primitive Society', we find that like Wittgenstein's writing it is reliant on second-hand ethnographic data. Consequently it is quite extraordinary that from his reading of Evans-Pritchard's study of the Azande, Winch fails to note and explain the unequivocal observation that:

Azande insist that magic must be proved efficacious if they are to employ it. They say that some magicians have better magic than others, and when they require a magician's services they choose one whose magic is known to be efficacious. . . . If a Zande is challenged and has to defend the virtue of his medicines, he does so by citing the occasions on which their potency has stood the test of experience. I could quote many examples.[56]

J. W. Cook draws attention to this observation and claims that it demonstrates Winch's argument to be 'defective from beginning to end'. For whilst it is true that:

when Azande cite evidence, it is often of the *post hoc ergo propter hoc* form . . . the question before us is not whether the Azande cite any evidence of the sort we would accept but only whether they cite evidence at all. And we can see from the foregoing passages that Winch is certainly wrong in maintaining that the Azande take no interest in evidence regarding their magic.[57]

---

[56] Cited by J. W. Cook, 'Magic, Witchcraft, and Science', *Philosophical Investigations*, 6 (1983), 15.
[57] Ibid. 16.

The case against intellectualism is dependent on the test of evidence and experience. In relation to the intellectualist interpretation of Zande magic we find that the evidence noted by Evans-Pritchard undermines the expressivist criticism of it, and so shows that the intellectualist interpretation makes better sense of this particular practice. Of course, that an expressive interpretation is unsatisfactory in this instance does not imply that it will generally be so, but it may be that the disregard for evidence concerning the behaviour of believers which we have discovered here is to be found in the treatment of Christian belief too.

7. The Case of Religion

The challenge of the expressive theory has been put in these terms: that the realist mischaracterizes the nature of religious language in seeking to justify it. I suggested in sections 3 and 4 of this chapter that Wittgenstein's opposition to an intellectualist account of magic and religion is to be construed as resting on the following argument: though the language of these practices is not manifestly expressive, the behaviour associated with it is rendered intelligible only by this interpretation. Of primary significance is the alleged complacency of believers towards what would be evidence if their beliefs were putatively fact-stating. This and similar behaviour favours the claim that such language is expressive. The same conclusion is allegedly reached if we apply to Christian belief the weak principle of charity which dictates that we maximize intelligible behaviour. Since believers do not treat their beliefs as they should treat hypotheses (they ignore the evidence), it is best to see them as expressing what goes deep in their lives, rather than as asserting some antique cosmology.

That there is a prima-facie presumption in favour of this interpretation need not be doubted (assuming that the evidence is as supposed). If the attitude of believers to evidence were agreed to be a thorough, uncaring disregard, then though the intellectualist account remains logically possible (as was made clear in discussing Wittgenstein's arguments against Frazer), the expressive interpretation does seem to be better. Maybe believers really do take their beliefs to be fact-stating and descriptive, and are too stupid to see how the evidence threatens their beliefs. In any particular case lack of concern for evidence may be explained away in this fashion. But it is clear that if the facts are as alleged, the onus of proof shifts to those who deny that religious beliefs are expressive.

However, it is this alleged lack of concern for evidence which has been challenged in the case of Zande magic and ought to be challenged in relation to religious belief. For if these beliefs are really expressions of feeling or responses to the way the world is rather than statements capable of truth or falsity, there would be no taking account of the evidence. But to maintain this leads, I shall argue, to a strained and unnatural account of the phenomenology of religious belief as we find it in Christianity. That this is so may escape the notice of some philosophers because they understand 'taking account of the evidence' in a naïve, falsificationist sense, and see the relationship of belief and evidence as extremely direct, straightforward and unproblematic. Viewed in that light, believers often do not 'take account of the evidence'. It is true, for example, that most believers do not worry about, let us say, the conceptual problems concerning continuity of identity through death and consequently do not scan the philosophy journals month by month to assess the current state of play. This important feature of religious belief, which I shall need in the next chapter to accommodate, gives the expressive position a certain plausibility, for its greatest strength lies in its promise to protect the secure status of religious belief in the believer's life and our discomfort in calling these beliefs 'hypotheses'. However, I shall argue that it achieves this at the cost of mischaracterizing other aspects of religious behaviour. In particular, explaining our discomfort in terms of the expressive theory renders mysterious much that we associate with loss of faith. It remains for the next chapter to defend the claim that a predominantly intellectualist understanding is able to balance both these elements in its own analysis of religious belief.

An obvious feature of the religious life is the desertion of it. This is a phenomenon too important to be ignored by any account of religious belief which claims to be descriptive. How is it to be explained? How does it happen that people fall away from religious faith? The intellectualist suggests that one of the causes of this is that beliefs which were once thought of as truly descriptive of the world and explanatory of its nature are found to be no longer credible as evidence impinges upon them.

What is the advocate of an expressive account to say of the fact of loss of belief? Phillips attends to the issue in his paper 'Religious Beliefs and Language-Games', and returns to it in 'Belief, Change and Forms of Life: The Confusions of Externalism and Internalism'.[58] The problem as Phillips sees it is that the expressive account can make it look as though the religious way is practically self-contained. But then, he asks, 'If religious beliefs were isolated language-games, cut off from everything which is not formally religious, how could there be any of the characteristic difficulties connected with religious belief?'[59] The problem has a corollary— just as the difficulties with religious belief seem to be lost if it is unrelated to the world, so too is its characteristic importance. Phillips provides an admirable statement of the difficulty when he explains the misgivings he has felt about the possible implications of the language-game analogy:

if religious beliefs are isolated, self-sufficient language-games, it becomes difficult to explain why people *should* cherish religious beliefs in the way they do. On the view suggested, religious beliefs seem more like esoteric games, enjoyed by the initiates no doubt, but of little significance outside the internal formalities of their activities. Religious activities begin to look like hobbies; something with which men occupy themselves at weekends.[60]

Whilst ridding us of doubts on this count, Phillips wants to avoid the trap of showing how religious beliefs can be justified, for this would involve a mischaracterization of them. He intends to prove that far from being cut off from the rest of our lives, religious beliefs as he understands them are intimately caught up with those things about which they speak, 'birth, death, joy, misery, despair, hope, fortune and misfortune'.[61] Presumably demonstrating how they are so caught up will demonstrate how they may be destroyed, and why they have the characteristic importance which we associate with them.

How, then, is the contingency and importance of religious belief to be reconciled with its expressive status? Take, for example, the problem of evil which is often found to be a grave test of a person's faith. Through a vivid experience of evil in his or her own life,

---

[58] D. Z. Phillips, 'Religious Beliefs and Language-Games', in *Philosophy of Religion*, ed. Mitchell, 121–42; and 'Belief, Change and Forms of Life: The Confusions of Externalism and Internalism', in *The Autonomy of Religious Belief*, ed. F. Crosson (Notre Dame, Ind.: University of Notre Dame Press, 1981), 60–92.
[59] 'Religious Beliefs and Language-Games', 135–6.
[60] Ibid. 122.   [61] Ibid. 134.

someone may come to see the world as irredeemably evil and lose his or her faith. Phillips recognizes the need to capture this phenomenon if his account of belief is to be convincing and offers the following as an explanation of how one might be unable to sustain an affirmation of God's love for the world on seeing the death of one's child:

One might want to believe, but one simply cannot. This is not because a hypothesis has been assessed or a theory tested, and found wanting. It would be nearer the truth to say that a person cannot bring himself to react in a certain way, he has no use for a certain picture of the situation.[62]

Phillips's suggestion is that the believer, having no more use for a certain picture, has been won over by a rival. But it is hard to comprehend this 'winning over' if it is not a succumbing to the force of evidence and argument. If the experience of a child's death is *not* related to the picture as evidence on which inference is based—as Phillips holds in saying that a hypothesis has not been tested and found wanting—then it could surely only be a confusion on the part of the believer to give up this picture. If faith, like prayer, is an expression of hope for one's life, why give up hoping? It is true that if hope takes the form of optimism, its being abandoned is explicable; but then that is just because a certain belief (namely that things will turn out well) has been found wanting in the face of the evidence. Phillips's account is peculiarly unconvincing here.

Whether or not Phillips senses doubts such as these, he returns to the problem of loss of faith in a recent paper. What he has to say, however, is of little help, for he gives an explanation of loss of faith which is wholly unrelated to the perspective of the believer. Once more he is trying to represent religious beliefs as they are, and he seeks a passage between the dangers of what he calls 'internalism' and 'externalism'. Externalism involves what he regards as the classical error of philosophy of religion which supposes that the meaning of religious discourse must be similar to the meaning of some other sort of language, such as scientific; internalism, the error to which its critics have thought that 'Wittgensteinian fideism' is especially prone, of so construing the meaning of religious language that it seems to be set apart from beliefs about the world. Phillips's 'middle way' aims to emphasize the distinctiveness of religion at the same time as demonstrating its connection with the

[62] Ibid. 136.

world. But what of the doubts which were outstanding in spite of his previous paper? Can he give an account of loss of faith which is not superficial and unconvincing?

As an example of how the world affects religious beliefs, Phillips tells of a harvest dance where

> what happens to the harvest may affect what happens to the dance. So although no account of the dance can be given in terms of technological causality, such technology may still affect the dance. The values of technology may erode values inherent in the dance. If the values of technology become dominant and all-pervasive, the dance will come to be regarded as a waste of time.[63]

Phillips's point is that the spread of technology may, for example, elevate efficiency above other concerns, such as for comradeship between workers. Obsession with the former may destroy the values expressed in the wish for a dance.

This may provide a convincing account of the abandonment of the imaginary harvest dance. It may also serve as an explanation of the erosion of an admittedly expressive religious practice, such as that of accompanying the dead to their graves in solemn procession. For if people are captivated by the notion of 'usefulness' and come to regard it as the sole criterion for judging actions, then practices expressive of respect may soon come to seem pointless and be forsaken. But as an account of the loss of faith which may accompany the death of a child it seems unpromising and highly implausible. In relation to the harvest dance, the new generation comes to see it as a waste of time. But for people who lose their faith in the face of death, the mild expression 'waste of time' would not capture their changed attitude to religion. It is not that worship is now something which ranks a few places lower in their priorities than it once did. It is rather that it now has no place whatsoever. 'I'd go to the dance if I weren't so busy' is just the sort of thing someone might say; so too, 'I haven't the time to lay flowers on the grave'. But 'I'd go and worship if I weren't so busy' is not how the parents who have lost their faith after the death of their child would explain their absence from church. There is no similarity between the abandoning of the harvest dance and the loss of faith.

Nor does Phillips's analogy cope any better with a real example, such as the crisis of faith which geology induced in the minds of

---

[63] 'Belief, Change and Forms of Life', 64.

many Victorians, such as Tennyson. *In Memoriam* describes a struggle to come to terms with the death of Arthur Hallam, at a time when science seemed to be threatening religious faith and the consolation it might provide. Tennyson had learnt from geologists such as Lyell that:

> The hills are shadows, and they flow
> From form to form, and nothing stands;
> They melt like mist, the solid lands,
> Like clouds they shape themselves and go.[64]

And in the light of Nature's inconstancy he asks:

> And he, shall he,
> Man, her last work, who seem'd so fair,
> Such splendid purpose in his eyes,
> Who roll'd the psalm to wintry skies,
> Who built him fanes of fruitless prayer,
>
> Who trusted God was love indeed
> And love Creation's final law—
> Tho' Nature, red in tooth and claw
> With ravine, shriek'd against his creed—
>
> Who loved, who suffer'd countless ills,
> Who battled for the True, the Just,
> Be blown about the desert dust,
> Or seal'd within the iron hills?[65]

It was not that the values of geology ousted religious values, rather that facts revealed about the earth's history threw into doubt traditional religious claims. As in the case of the problem of evil, the religious believer does not think of loss of faith as a change of attitude, but instead experiences the hardness and immediacy of a direct factual challenge.

Phillips's account of loss of faith in religion is inadequate. It misses the perspective of the believer whose faith has been challenged by evil or by some scientific discovery, and who is led to say, 'I could no longer believe in God when my child died', or 'After studying cosmology I have given up belief in divine creation'. We might highlight the inadequacy by contrasting the response of a

---

[64] *In Memoriam*, cxxiii. Tennyson was so taken up with the geological controversies that he trembled as [he] cut the leaves' of Chambers's *Vestiges of Creation*. Quoted in P. Turner, *Tennyson* (London: Routledge & Kegan Paul, 1976), 124.

[65] Ibid. lvi.

religious believer with that of a fatalist. The latter's attitude is not affected by events in the world for nothing can compel the abandonment of fatalism. Now according to Phillips the person of faith is similar to the fatalist. For those who pray express their sense of their weakness in the face of the difficulties of living up to their ideals and their deep hope for persistence in moral endeavour, and adopt a reflective attitude to life. Similarly, those who believe in eternal life express the regard in which they continue to hold the deceased. Given these descriptions, distinguishing the believer from the fatalist becomes impossible. The fatalist says 'it can't be helped' or 'that's the way it goes', and just as no event in the world can affect in a direct way the expression of this attitude so also the fatalist does not possess a *reason* for expressing it. It is true, of course, that the fatalist might give up saying these things, since 'that's the way it goes' may be appropriate in relation to dropping an ice-cream in the sand, but not in the face of the death of one's child. Yet the giving up of such a happy-go-lucky attitude is not *necessary* from this point of view, perhaps just seemly.

But in the case of religious belief we cannot imagine that loss of faith is thought 'seemly' by those who experience it. This is not how it is with the religious person. The problem of evil, for example, and perhaps an awareness of certain scientific advances, may cause for the erstwhile believer a painful loss of faith which is experienced as a loss of reasons. It is not that believers no longer have any use for a picture, but that they feel compelled no longer to use it.

Phillips threatens to assimilate believer and fatalist, whereas the trauma of loss of faith or the sense that modern science has somehow undercut one's beliefs is quite unknown to the fatalist for whom any clash between a non-cognitive fatalism and scientific discoveries or events in the world is illusory. Hence, in spite of the genuine insights in Phillips's account of religious language and the ingenuity he displays in attempting to mitigate the problems it has in explaining the difficulties of faith, there is an ultimate superficiality about his account when it seeks to explain the challenge of evil and science to religious belief.

The genuine insight in the expressive understanding of religious belief lies in its stressing the security of those beliefs in the believer's life. It is a security which seems to be greater than an account of these beliefs as 'hypotheses' would warrant. If we are to look at what people do in trying to see what they mean, as the weak principle of charity maintains we should, then we cannot neglect this feature of religious faith. I have tried to suggest, however, that in its concern to give protection to the fact that beliefs are relatively secure, the expressive thesis renders entirely mysterious other aspects of religious behaviour. Above all, Phillips's account seems to do violence to the facts of loss of faith as we naturally experience them.

It remains to be seen whether the intellectualist is able to cope with these two poles any better—whether, that is, it is possible that religious belief which rests on an explanatory justification, and hence an assessment of evidence, may yet display the tenacity which is characteristic of it. Until it is shown that it is possible for the intellectualist to take seriously the relative immunity of religious belief to criticism, Phillips might well claim that the intellectualist is, at best, some way from a proper understanding of the logic and status of religious discourse.

5.

Faith and the Religious Adequacy of Explanatory Justification

1. Introduction

THE last chapter was concerned with the appropriateness of an explanatory justification of Christian theism to religious faith. Wittgenstein and Phillips expound the thesis that religious discourse is not fact-stating, from which it follows that the sort of justification envisaged is irrelevant, for it rests on a confusion as to the nature of such discourse. This sort of justification might be appropriate to religious belief if it were descriptive, but it is not.

I have argued that the thesis that religious belief is expressive rests primarily on its claim to render intelligible the alleged indifference of believers to what would be evidence were their beliefs factual. At the end of the last chapter, however, I have noted that such a means of coping with the security of religious belief proves too drastic, for it fits ill with other aspects of the life and attitudes of the believer where evidence does seem important. None the less it remains to be shown that an account of religious language as cognitive is able to protect the insight that these beliefs are, in a sense, resistant to criticism.

The argument put forward by Wittgenstein and Phillips would show, were it successful, that an explanatory defence is religiously irrelevant; other arguments, which also note the resistance to evidence of religious beliefs, seek to show that it is religiously inadequate. It is not that the justification mistakes the logical type of religious language, but rather that it fails to protect other key features of religious belief. For surely, it will be said, religious faith must have certain characteristics and these characteristics would be inimical to faith which rested on an explanatory justification.

I shall consider three arguments of this sort against the religious adequacy of explanatory justification, arguments which can be attributed to Lyas, Newman, and MacIntyre. I shall claim that the first argument makes illicit use of the notion of a belief as 'religiously adequate'; that the second objection can be dealt with by attention to the meaning of the word 'belief'; and that the third demands a demonstration that confidence is possible in a faith founded on a cumulative case. It is in answering this last demand that our duty to Phillips and Wittgenstein is finally discharged.

2. An Argument from the Motive for Faith

Faith which rested on an explanatory justification would be religiously inadequate in that it would not satisfy the motive from which religious belief arises. Thus argues Colin Lyas in an article entitled 'The Groundlessness of Religious Belief'.[1] The motive for religion, so Lyas suggests, is a sense of one's contingency and a consequent desire for security. What makes a response to a situation religious is that it is resolved by a sense of safety coming from outside the world, not dependent on anything which happens within it. Lyas contends:

From the source of religion that I have tried to indicate there arises also the requirement that the religious belief that is to sustain a person in need of it must not be a belief that depends on what may happen in the world or turn out to be the case. It is this which gives some sense to the claim that a *religious* belief must not be one that depends upon and changes with *evidence*.[2]

Why *must* it not depend on 'what may happen in the world or turn out to be the case'? To appreciate the force of this 'must' Lyas suggests that we consider what it is to believe in something on the basis of evidence. Take our belief in the existence of the planet Pluto. Two things should be noted, he thinks. First, the belief is in the probable existence of an object, and what is probably true is possibly false. Second, observations may be made which would undermine the belief in the existence of the object so that the belief would rise and fall with the evidence.[3] Hence it is clear that the believer's belief in God could not be an evidenced belief in the existence of something, for:

If his belief thus depends on evidence, his peace is still at the mercy of what

[1] C. Lyas, 'The Groundlessness of Religious Belief', in *Reason and Religion*, ed. S. C. Brown (London: Cornell University Press, 1977), 158–80.
[2] Ibid. 174–5.
[3] Ibid. 175.

might happen. For other evidence could turn up. If, however, he *is* still at the mercy of what might happen when he believes in God then that belief in God is not adequate to the religious need from which it arises. *That* sort of God is not a religiously adequate God. Belief in it cannot make us safe from what might happen.[4]

A belief in a religiously adequate God cannot be a belief based on evidence, for 'as *we* use the term religion, a religious belief in God could not rise or fall with the evidence.'[5]

There is something more than slightly fishy about this argument. Imagine that someone said that a sound investment is one made without reference to the state of the market. Now how might the assumption that all sound investments are groundless be defended? One way would be to show that the motive for investment arose from a sense of one's financial insecurity and from a desire to be safe. It follows that any investment made on the basis of assessments of the market would not satisfy the motive of investment, since the market could rise or fall. Therefore we have a proof that a sound investment is not made according to the state of the market.

The fishiness of Lyas's a priorism comes to this: that the underlying motive for an endeavour (be it religious or financial) will not always be satisfied by its results. So even if Lyas is right about the motive for religious belief, nothing follows. A demand for financial security may not be capable of realization in the market-place. The search for peace may be the motive for an interest in religion, but may not be the reward of the search. For the fact that we are looking for something is no guarantee that we will find it, and in relation to religion in particular, we should be aware of the possibility that the seeker may come to see that what has been found is not identical with the original object of the search.

In Lyas's hands the term 'religiously adequate' is wrongly used. Faith can indeed be religiously inadequate, but when we say so we do not usually mean that it is such in relation to some postulated motive for belief. If we did so speak, then what is religiously adequate for me might not be so for you, in which case 'religiously adequate' has come to mean little more than 'satisfying'. Now Lyas may or may not be right that belief which might fall prey to

---

[4] 'The Groundlessness of Religious Belief', 175–6.
[5] Ibid. 178.

evidence is unsatisfying, but that does not seem to me to be an important point. A more serious charge conveyed by the words 'religiously inadequate' would be that one's belief is wanting because it falls short of the demands which the religious way itself imposes. Faith may be wanting, or religiously inadequate if it fails to match up to a paradigm of faith which is characteristic of the tradition and normative for it. This charge is a much more dangerous threat to the interpretation of theism as based on inference to the best explanation than the one made by Lyas, for what discussion of this point really demonstrates is the irrelevance of the motive for faith in establishing what faith should be.

3. The Nature of Faith

The essential error in Lyas's argument lies in supposing that one can establish the nature of religious faith from the alleged motive for such belief. On the contrary, even if one correctly diagnoses the motive for faith, its character is still left in doubt. Hence Lyas has not shown that faith could not be founded upon the explanatory justification of religious belief which the theist is encouraged to advance by reflection on the currently favoured defence of scientific realism.

A more telling objection is one suggested by certain remarks made by Newman, and it takes the form of noting a characteristic of faith which, it is claimed, would not be characteristic of faith defended as explanatory. It is maintained that faith which rested on an explanatory justification would be religiously inadequate in that it would not involve belief. For whilst the proper epistemic attitude involved in religious faith is indeed belief, the attitude which we adopt towards, say, a scientific hypothesis justified in the manner outlined in Chapter 3, is not belief. In other words: belief is a necessary (if not sufficient) condition for Christian faith; explanatory justification is not a justification of belief; hence an explanatory justification of faith is religiously inadequate.

The attribution of such a criticism to Newman is warranted by an interesting amendment he made to an opinion he had expressed as an Anglican, when he was inclined to characterize faith by reference to a readiness to act on the propositions of the religious outlook. Faith consists not in being persuaded or certain of the truth of these propositions (that is, not in possessing a particular epistemic attitude towards them); rather faith consists in a commitment or readiness to act upon the assumption of their truth.[6] Thus in *Via Media of the Anglican Church*, Newman proposed that 'Action is the criterion of true faith' and wrote that our behaviour constitutes an instance of faith when 'even the confused cry of fire at night rouses us from our beds'.[7]

As a Roman Catholic some years later, Newman corrected his early works. His former view was too lenient in that faith defined as a readiness to act seems to require no cognitive attitude at all. And this leniency is surely wrong. Take a proposition basic to Christian theism, namely, that God exists. The denial of this proposition seems to be incompatible with faith, for the confession 'I have faith in God, though I don't believe that he exists' is not the confession of someone who exemplifies the Christian virtue of faith, even if it is accompanied by the sort of actions appropriate to one who does in fact believe.

Thus Newman was right to abandon his earlier views, in so far as they seemed to imply that someone who merely acts as if God exists exemplifies the virtue of faith; there must be some stronger epistemic requirement. But his later view moves towards too rigorous a standard, Newman now characterizing faith by means of a distinction between practical and speculative certitude.[8] Practical certitude is the product of evidence sufficient to convince us to act in a certain way, speculative certitude the result of evidence leading to an unhesitating and confident judgement that a belief is true. Newman came to think that the former was not good enough for faith (as it is not, for the mere possibility of a proposition's being true may be sufficient to convince us to act as if it is), and in amending the earlier remark he wrote, 'No one would say we believed our house was on fire, because we thought it safest on a cry of fire, to act as if it was.' Instead, he wrote, 'By belief of a thing this writer understands an inward conviction of its truth.'[9]

Newman held not only that speculative certainty was essential for faith, but that where one lacked it, one was not properly

[6] See W. R. Fey, *Faith and Doubt: The Unfolding of Newman's Thought on Certainty* (Sheperdstown, W. Va: Patmos Press, 1976), 33.

[7] Cited by Fey, ibid., 33.

[8] See M. J. Ferreira, *Doubt and Religious Commitment: The Role of Will in Newman's Thought* (Oxford: Clarendon Press, 1980), 47–51.

[9] Notes added to *Via Media* in 1877, cited in Fey, *Faith and Doubt*, 34.

described as believing the proposition in question—such as 'that this house is on fire'. Thus Newman's later understanding of this issue was that there is only a case of belief where we are possessed of certainty, marked by an unhesitating and confident judgement that a proposition is true. Since this is generally not the case with a proposition which is chosen for its explanatory power—where it may be that our judgement is that the proposition is probably true—it follows that explanatory argument does not justify belief. However, Newman's reaction against his earlier leniency is too strong and results in too restricted a view. That it has little to commend it as an analysis of our usage of the word 'believe' becomes clear in Swinburne's discussion of the nature of belief.[10]

Swinburne proposes that 'normally to believe that p is to believe that p is probable', and as a categorization of our terms that the following three cases are all cases of belief:[11]

(1) I believe p where I know p to be true. This seems to be straightforward.

(2) I believe p where I believe p to be more probable than not-p. That is to say, if I think that it is more likely than not that p, then it follows that I believe it.

(3) I believe p where I believe p to be more probable than any one of the alternatives; that is, where there is more than one. So, I may not think it likely (having a chance greater than 1 in 2), that p, but given the range of options, I believe p to be more probable than any of its alternatives. I may not think that it is likely that Liverpool will win the Cup, but if I think that Liverpool are the team most likely to win the Cup, then in reply to the question 'who do you believe will win the Cup?' my natural response would be 'Liverpool' rather than 'I do not believe of any particular team that it will win.'[12]

Swinburne thinks that these cases should be distinguished from another, which we are inclined to deny to be an instance of belief, where one acts on the assumption that p. Such a description most naturally fits those cases where one behaves in a way appropriate to

[10] Swinburne, *Faith and Reason*, ch. 1.
[11] Ibid. 4. It should be noted that Swinburne does not think that his claims are analytic; he acknowledges that he may in fact be suggesting what would amount to a 'tidying up' of our concept of 'belief'. Even so, his account provides us with a more natural usage than Newman's.
[12] Ibid. 5.

the person who believes p, but without thinking that p is probable to any significant degree; that is, neither more probable than not-p, nor more probable than any of the alternatives. An example of acting-as-if is found in the behaviour of the agnostic described by Anthony Kenny at the end of *The God of the Philosophers*:

There is no reason why someone who is in doubt about the existence of God should not pray for help and guidance on this topic as in other matters. Some find something comic in the idea of an agnostic praying to a God whose existence he doubts. It is surely no more unreasonable than the act of a man adrift in the ocean, trapped in a cave, or stranded on a mountainside, who cries for help though he may never be heard or fires a signal which may never be seen.[13]

As Kenny suggests, behaving as if p were true may be rational in certain circumstances even though one doubts that p is probable.

Newman's suggestion seemed to be that a characteristic of 'believing' something is a confidence in its truth. If one does not have what he terms speculative certitude, one does not believe. Since faith involves belief, it follows that the so-called explanatory justification here mooted is religiously inadequate for it would not license such speculative certitude. What has been suggested in reply is that though Newman properly reacted against his earlier views, his later ones blur distinctions which we ought to draw. Armed with those distinctions we can see what is right in the point that Newman was trying to make, namely, that evidence sufficient to make us act as if a proposition is true is not necessarily enough to make us believe that proposition. But in reacting against that and claiming that it is only where we are certain of something that there is an instance of belief, all other cases being of 'acting-as-if', he misses further distinctions which must be made. Whilst we should distinguish acting-as-if from belief, it seems that we apply more lenient standards in judging something to be a belief than Newman allows. Hence if we are to show that explanatory justification is irrelevant to religious faith, we have to do more than appeal to our usage in relation to the word 'belief'. Belief warranted by an inference to the best explanation is not a certainty of truth, but it may be belief none the less.

---

[13] A. J. P. Kenny, *The God of the Philosophers* (Oxford: Clarendon Press, 1979), 129.

4. Religion and Decisive Adherence

Faith involves belief, and explanatory justification may be enough to establish belief. But even with so much accepted, another front on which to attack the envisaged approach opens up. Thus it might be conceded that explanatory justification would justify belief, but not the right sort of belief. For the question which remains outstanding is whether the faith of the religious believer, if it is to be religiously adequate, must somehow be more than mere belief.

This is the substance of an objection, related to the one suggested by Newman's writing, which says that an explanatory justification of faith would be religiously inadequate for the reason that such faith would be wavering. In other words there is a disparity between the character of true faith and the sort of belief which inference to the best explanation would justify. In particular, faith involves and is characterized by a certain sort of confidence. And this shows that the proposed defence is irrelevant for it would not warrant confidence at all. That is, though both faith and propositions defended by an explanatory justification may involve belief, the sort of belief demanded by the one is different in kind from the sort of belief licensed by the other. So though Newman was wrong to deny the status of 'belief' to attitudes which fall short of confidence of truth, he was right to draw attention to the fact that not just any belief is a proper foundation for faith.

Faith, after all, is markedly robust and full of confidence—at least in the paradigms which are held out to inspire believers. Take, for example, the portrait of those who exemplify the Christian virtue of faith presented in the Epistle to the Hebrews. The author tells us of those

who through faith conquered kingdoms, enforced justice, received promises, stopped the mouths of lions, quenched raging fire, escaped the edge of the sword, won strength out of weakness, became mighty in war, put foreign armies to flight. Women received back their dead by resurrection. Some were tortured, refusing to accept release, that they might rise again to a better life. Others suffered mocking and scourging and even chains and imprisonment. They were stoned, they were sawn in two, they were killed with the sword; they went about in skins of sheep and goats, destitute, afflicted, ill-treated—of whom the world was not worthy—wandering over deserts and mountains, and in dens and caves of the earth.[14]

Would faith be so robust if based on an explanatory justification? Such a justification might support some beliefs of a religious kind, but not religious faith as we know it. As Newman puts the point: 'Without certitude in religious faith there may be much decency of profession and of observance, but there can be no habit of prayer, and no directness of devotion, no intercourse with the unseen, no generosity of self-sacrifice.'[15]

Such doubts about the religious adequacy of explanatory justification are implicit in MacIntyre's objection to an argument of Ian Crombie's, which, he protests, makes of

religious belief . . . a hypothesis which will be confirmed or overthrown after death. But, if this is correct, in this present life religious beliefs could never be anything more than as yet unconfirmed hypotheses, warranting nothing more than a provisional and tentative adherence. But such adherence is completely uncharacteristic of religious belief. A God who could be believed in in this way would not be the God of Christian theism. For part of the content of Christian belief is that a decisive adherence has to be given to God. So that to hold Christian belief as a hypothesis would be to render it no longer Christian belief.[16]

MacIntyre's argument here seems to go as follows:

(1) an explanatory hypothesis is given only a provisional and tentative adherence.

(2) religious belief demands not provisional and tentative but decisive adherence.

(3) therefore religious faith cannot be founded on an explanatory hypothesis.

To judge of the truth of the conclusion we naturally have to devote some attention to the premises. We shall see that the strongest support they have is from each other, and that pressed they both lose some of their appeal.

If something not known to be true is held as a hypothesis, does that mean that our adherence to it will be provisional and tentative?

---

[14] Epistle to the Hebrews, 11:33–38, Revised Standard Version.

[15] J. H. Newman, *An Essay in Aid of a Grammar of Assent*, ed. I. T. Ker (Oxford: Clarendon Press, 1985; 1st publ. 1870), 144.

[16] A. MacIntyre, 'The Logical Status of Religious Belief', in *Metaphysical Beliefs*, ed. S. E. Toulmin, R. W. Hepburn and A. MacIntyre, 2nd edn. (London: SCM, 1970), 171.

This is the assumption made by MacIntyre and also the assumption implicit in Newman's contrast between 'decency of profession' and 'directness of devotion'. Certainly it has to be allowed to Newman and MacIntyre that if our faith is tentative and provisional, it will have none of the marks which go with devotion as we usually understand it. But the question is whether it is proper to suppose that all commitments to hypotheses *must* be tentative. Since discussion has so far focused on science, we may ask: first, 'is it true to say that in science hypotheses are held tentatively?'; and second, 'if they are not held tentatively, is there good reason why not?'

It was Sir Karl Popper who did much to foster the notion of hypotheses as held in an entirely tentative way, this being for him a requirement of the scientific quest.[17] As Lakatos puts it: '*Belief* may be a regrettably unavoidable biological weakness to be kept under the control of criticism: but *commitment* is for Popper an outright crime.'[18] Popper's portrait of the scientific enterprise takes its starting-point from his repudiation of induction, from which it follows that the past successes of a theory provide us with no assurance of its future adequacy. But if this is the case, the scientist's task cannot be to establish theories, nor a scientist's duty to believe them. Instead the scientist is supposed to propose some bold theory which tries to capture and explain all the available data, and to seek its refutation by deriving from it predictions which then face the tribunal of experience. When a refutation is discovered, a new theory is devised, itself subject to refutation. On and on the process goes, as science advances through the falsification of a succession of theories. The scientist holds on to the presently unfalsified theory and it might even be said that 'We tentatively "*accept*" this theory—but only in the sense that we

[17] Lakatos has distinguished three Poppers, the dogmatic, naïve, and sophisticated falsificationist, the first being a creation of his critics; see 'Criticism and the Methodology of Scientific Research Programmes', *Proceedings of the Aristotelian Society*, 69 (1968–9), 149–86. This should suffice to alert us to the exegetical difficulties associated with Popper's work, difficulties which are not, however, central to the present discussion. Suffice it to say that the account I give of Popper as a naïve falsificationist represents, according to Lakatos, a genuine element in Popper's thought which has, moreover, formed the popular impression of his message. Popper's major work is his *Logic of Scientific Discovery* (London: Hutchinson, 1959). See also his 'Science: Conjectures and Refutations' in *Conjectures and Refutations*, 4th edn. (London: Routledge & Kegan Paul, 1972), 33–65 and 'Conjectural Knowledge: My Solution of the Problem of Induction', *Objective Knowledge*, 2nd edn. (Oxford: Clarendon Press, 1979), 1–31.
[18] 'Falsification and the Methodology of Scientific Research Programmes', 92.

select it as worthy to be subjected to further criticism, and to the severest tests we can design.'[19]

Popper's rejection of induction renders the analogy between scientific theory and religious belief more incongruous than it might at first seem, for scientists are supposed not to *believe* theories which rely for their justification on inductive reasoning. But this aspect of the disanalogy we may disregard, for the simple reason that Popper has been unable to carry through an account of science which takes seriously the eschewal of reliance on inductive reasoning. Putnam asserts that 'The distinction between *knowledge* and *conjecture* does real work in our lives; Popper can maintain his extreme skepticism only because of his extreme tendency to regard theory as an end for itself.'[20] His point is that applications of theories are made not just for testing those theories, but for practical purposes; hence those applications anticipate the future successes of a theory. But why suppose that a theory which has withstood tests today will withstand them tomorrow, unless one makes an inductive inference about the future resembling the past?

Scientists believe in their theories; but even if we ignore the anti-inductivist strain in Popper's work which talks of theories as 'accepted' rather than as 'believed', the disanalogy between science and religion still stands. For if Popper is in part right about the pattern of conjecture and refutation, then there will be an absence of commitment in science, and an air of cool and detached reserve in the scientist's search for falsification which will be far removed from the passions and confidence of religious belief. The problem is not that the scientists do not believe in their theories—they do. Rather it is that they exhibit no attachment to them. In contrast, the religious believer does not switch theories at every opportunity. Faced with a difficulty in relation to a religious belief, the believer will usually not abandon it without a struggle; hence, if Popper is to be believed, any proposed analogy between scientific belief and faith looks quite ridiculous.[21]

---

[19] Popper, *The Logic of Scientific Discovery*, 419; cited by H. Putnam, 'The "Corroboration" of Theories', in *Philosophical Papers, i. Mathematics, Matter and Method* (Cambridge: Cambridge University Press, 1975), 253.

[20] 'The "Corroboration" of Theories', 252.

[21] For an example of a philosopher of religion relying uncritically on a Popperian view of science to establish another disanalogy, this time between the possibility of testing a theory and the possibility of testing prayer, see V. Brümmer, *What Are We Doing When We Pray?* (London: SCM, 1984), 1–7.

Popper's account of the scientist's behaviour makes it nothing like that of the person who has faith, and so supports the view that a defence of a religious belief which is analogous to the defence of a scientific theory will be religiously inadequate. It would thus appear to vindicate MacIntyre's first premiss, that an explanatory hypothesis will be given only tentative adherence. Kuhn and Lakatos, however, have brought attention to the fact that Popper's methodology is neither an accurate portrait of how an hypothesis functions in science, nor a rational proposal for how scientists might improve their discipline. Kuhn writes that 'No process yet disclosed by the historical study of scientific development at all resembles the methodological stereotype of falsification by direct comparison with nature.'[22] And elsewhere he claims:

All experiments can be challenged, either as to their relevance or to their accuracy. All theories can be modified by a variety of *ad hoc* adjustments without ceasing to be, in their main lines, the same theories. It is important, furthermore, that this should be so, for it is often by challenging observations or adjusting theories that scientific knowledge grows. Challenges and adjustments are a standard part of normal research in empirical science, and adjustments, at least, play a dominant role in informal mathematics as well.[23]

The problem with Popper's theory of conjecture and refutation is that it proposes too simple a relationship between theory and world. It is as if, for Popper, the facts which face the theorist are always firm and unchallengeable, so that, in Lakatos's words, science is a 'two-cornered fight' between experiment and theory.[24] But suppose that some theory meets with several 'facts' which seem to endanger it, as did Newton's theory of universal gravitation. It was supposed to predict the positions of the planets, but did so with only a fair degree of success. One famous anomaly was its prediction of the moon's perigee; that is, the time at which the moon makes its closest approach to the earth during its orbit. Newtonian predictions were significantly wrong. According to Popper, Newton should have thrown up his hands, accepted the refutation of his theory and started all over again. But there are many options other than the Popperian one of going back to the drawing board. Let me give three illustrations.

[22] *Structure of Scientific Revolutions*, 77.
[23] 'Logic of Discovery or Psychology of Research?', 13.
[24] 'Falsification and the Methodology of Scientific Research Programmes', 115.

First, one may assert that the 'facts' are false, and seek for some explanation—such as inaccurate experimental procedures—which will account for the unfavourable findings. For it is perhaps no more than bad measurement which accounts for the divergence between what is observed and what is predicted. On a number of occasions when Newtonians were faced with unfavourable findings they confidently told the observers to check their methods of observation and measurement, and on a number of occasions the observers had to admit that the theory had been right, and their observations wrong.[25]

Second, one may look to the actual details of the application of one's theory; perhaps the 'anomalous' observation really is entailed by one's theory after all. This occurred in the case of the observations of the behaviour of the moon's perigee, which seemed (from a Newtonian perspective) to be anomalous; this phenomenon fitted back into the theory with Clairaut's demonstration that the mathematics of the application had been wrong.[26] That this sort of error may have occurred when theory and observation clash should hardly cause surprise, for as theories become more and more complex the derivation of predictions from them becomes similarly complex. Mistakes can be made, and their discovery difficult. So even resistance to change has a value, as Kuhn maintains: 'By ensuring that the paradigm will not be too easily surrendered, resistance guarantees that scientists will not be lightly distracted and that the anomalies that lead to paradigm change will penetrate existing knowledge to the core.'[27]

Third, one might propose a revision of the auxiliary statements or initial conditions which, in combination with the theory in question, have allowed the derivation of predictions. If a planet does not appear to be on its predicted course then this may be accounted for by the presence of another, as yet undetected, planet. Thus the Newtonians accounted for the behaviour of Neptune.

[25] See Lakatos's 'classical example of this pattern' in Newton's dealings with Flamsteed, the Astronomer Royal; 'Falsification and the Methodology of Scientific Research Programmes', 130 n. 5.

[26] Kuhn, *Structure of Scientific Revolutions*, 81.

[27] Ibid. 65. Basil Mitchell examines the dangers for all academic disciplines in being 'lightly distracted' in his 'Faith and Reason: A False Antithesis?', *Religious Studies*, 16 (1980), 131–44. Mitchell captures that danger in an earlier work with his remark: 'The man who does not believe in anything with sufficient seriousness and consistency to test it in his life may make no mistakes; but he will not learn any lessons.' *Justification of Religious Belief*, 132.

Summing up the history of the Newtonian gravitational theory, 'possibly the most successful research programme ever', Lakatos writes:

When it was first produced, it was submerged in an ocean of 'anomalies' (or, if you wish, 'counter-examples'), and opposed by the observational theories supporting these anomalies. But the Newtonians turned, with brilliant tenacity and ingenuity, one counter-instance after another into corroborating instances, primarily by overthrowing the original observational theories in the light of which this 'contrary evidence' was established. In the process they themselves produced new counter-examples which they again resolved.[28]

Of the Newtonians Lakatos concludes with Laplace that they 'turned each new difficulty into a victory of their programme'.

The important point is that in cases such as these there is no simple relationship between conjecture and refutation, not least because there is not the simple theory–prediction link which Popper supposes.[29] So whilst for Popper anomalies constitute falsifications, for Kuhn and Lakatos, who see more clearly the complexity of the relationship between theory and prediction, they may be treated as problems or puzzles. That is to say, that though there may be no immediate, specific, or successful attempt to restore consistency to the situation, during a period of what Kuhn would term 'normal science'—which is to say during the period of the acceptance of a particular paradigm—prima-facie anomalies are treated not as refuting the paradigm, but as posing problems for which solutions will be forthcoming.[30] Kuhns's insight here, already mentioned in Chapter 2, is that a refutation of a theory does not consist solely in the discovery of a phenomenon which it cannot handle. He writes that 'if an anomaly is to evoke crisis, it must usually be more than just an anomaly.'[31] A striking example of this truth, which

---

[28] 'Falsification and the Methodology of Scientific Research Programmes', 133. Lakatos's other favoured example of a successful research programme born in a sea of anomalies is Prout's theory of 1815 that the atomic weights of all chemical elements were exact multiples of the weight of the hydrogen atom. This theory was one which went through several unrewarding periods for researchers ('degenerating problem shifts', as Lakatos terms them), before it emerged to win the day nearly a century later.

[29] Putnam sums up his argument to this effect in the claim that 'theories are *not* strongly falsifiable'. See 'The "Corroboration" of Theories', 258.

[30] *Structure of Scientific Revolutions*, 77 f.

[31] Ibid. 82.

highlights the inadequacy of the simple model of 'conjecture and refutation', is provided by Lakatos. He refers to the anomalous behaviour of Mercury's perihelion which was known for decades. However, 'only the fact that Einstein's theory explained it better transformed a dull anomaly into a brilliant "refutation" of Newton's research programme.'[32]

In place of the simplicity of Popper's scheme we shall need a more sophisticated account of the refutation of theories. Such an account would note that an anomaly is normally counted not as a falsification of the theory which constitutes the prevailing paradigm, but as an indication, for example, that the initial conditions assumed in the derivation of predictions must be revised. On the subject as to when normal science becomes extraordinary science (that is, when anomalies are taken as refutations) Kuhn is vague, insisting, above all, that there is no 'sharp dividing line' between puzzles and counter-instances.[33]

Lakatos proposes to capture the facts concerning the response to anomaly in a logic of scientific discovery which will amount to a sophisticated falsificationism. It is not individual theories which face the world he suggests, but research programmes. A research programme is individuated by the contents of its 'hard core', those central postulates of the programme which are beyond challenge from the point of view of its proponents. If those postulates are promising then the reaction to anomalies will be that outlined above: question the auxiliary belt of assumptions which, together with the hard core, has led to the anomalies. And if this policy pays off, it will be with increasing confidence that the anomalies are turned aside. The expectation will develop that a threatened anomaly is just one more puzzle to be solved.

W. H. Austin has tried to make use of Lakatos's classification of the logical structure of scientific theories in a comparison with the nature and structure of religious doctrines.[34] Lakatos's distinction between the hard core of a research programme and its auxiliary belt, is carried over to the religious case, where there are certain doctrines upon which the believer places much reliance, and others

[32] 'Falsification and the Methodology of Scientific Research Programmes', 159.
[33] *Structure of Scientific Revolutions*, 80.
[34] W. H. Austin, *The Relevance of Natural Science to Theology* (London: Macmillan, 1976), 111–12; and 'Religious Commitment and the Logical Status of Doctrines', *Religious Studies*, 9 (1973), 39–48.

which are relatively peripheral to the theistic scheme. These central doctrines constitute the hard core, and as against the auxiliary hypotheses are held in a way which is neither provisional nor tentative.

It is, I think, unprofitable to follow Austin with Lakatos down this particular path. There is no harm in taking over Lakatos's terminology, so long as one is aware that the sharp division he seeks to draw between hard core and auxiliary belt is not as easily made as he imagines. But a more positive reason against a close identification with Lakatos's analysis lies in his employment of this apparatus in the elaboration of an account of the confirmation of theories, which is, so I shall maintain in the next chapter, seriously flawed.[35]

The historical findings of Lakatos's work which support the rejection of naïve falsificationism may be accepted without endorsing the attempt he made to express those findings in a logic of scientific progress, for the two are clearly distinct. For whether or not one follows Lakatos in the details of his methodology, he and Kuhn provide us with a useful catalogue of examples where Popper's portrait of scientific activity is in direct opposition to the way science has actually worked. Instead of scientists going about their business by rejecting theories whenever findings seem unfavourable, we note them rejecting evidence, or going back to their calculations to see whether they might not come out better a second time. Instead of scientists who do not believe their theories and who are not committed to them, we find scientists so confident about them that they tell experimenters that they have got the facts wrong, and astronomers that they happen to have missed some planet roving around the heavens undetected. So much, then, for conjecture and refutation.

Is this tenacity and assurance a legitimate attitude to adopt towards a paradigm and is Popper's methodological directive to the scientist mistaken? It would seem so, for Popper's simple account of conjecture and refutation makes no allowance for the complexity of the relationship between theory and evidence. Because of this complexity a theory which shows promise must be allowed time to develop and to cope with the difficulties which it encounters. That Popper's advice to the scientist would have entailed the rapid

[35] For some of the problems which Lakatos's theory faces, see Newton-Smith, *Rationality of Science*, ch. 4.

rejection of the Newtonian paradigm is rightly taken by his critics as a *reductio* of his position. If a hypothesis holds promise of explanatory power then it is rational to allow it time to deal with the problems which confront it, for the explication of a rich idea is not likely to be a simple matter. Further, our attitude towards a theory will rightly be one of increasing confidence as the theory develops into a more satisfying explanation of the facts, and as more of the anomalies which challenge it are dispatched. Scientific theories often merit more than tentative adherence.[36]

MacIntyre is shown, therefore, to be mistaken in the first premiss of his argument, be it understood either descriptively or normatively, when it asserts that a hypothesis is always held tentatively. Some are. But not high-level ones where the success of an explanatory hypothesis provides reason for confidence in its continued power. The confusion between the levels is common but Mitchell insists:

the individual scientist's situation in relation to the body of accepted scientific knowledge is quite different from his situation in relation to a particular hypothesis, which he has formulated and is trying to test. Such a hypothesis he will indeed be prepared to accept or reject in accordance with experimental findings, when suitably repeated by himself and others. He can, so to speak, take it or leave it: and whichever he does, in normal cases, will not profoundly affect the structure of his science. This is the situation we generally have in mind when we think of someone treating something as a hypothesis. Since it represents the normal situation of the research scientist it is natural to take it as the paradigm of the scientific attitude. But the scientist cannot, and should not, adopt this 'take it or leave it' attitude to the main body of, e.g., physics.[37]

So whilst commitment to every scientific hypothesis is in a sense provisional (in that it depends on what happens in the world) it is not thereby shown to be tentative, if by that is meant 'held with no confidence or commitment'.

The second premiss of MacIntyre's argument asserts that religious belief demands a 'decisive adherence'. As I have said already, this claim gains its greatest plausibility from the false options presented by the assertion of his first premiss (see above, 104). Faced with a choice between saying that religious believers

---

[36] A more developed answer to the question 'when do they merit this adherence?' will be sought in the next two chapters.

[37] *Justification of Religious Belief*, 123–4.

have a tentative belief or give a decisive adherence, the obvious falsity of the first option drives us into the arms of the second. Fully aware of the choices open to us—aware, that is, that tentative adherence is not characteristic of science and hence not the only option as against decisive adherence—and informed of what 'decisive adherence' is taken to mean, we may be less inclined to accept that second premiss. What is required then, is that we should look at the second step in MacIntyre's argument in its own right. Is it true that religious belief demands decisive adherence?

What does this claim actually mean? MacIntyre treats 'decisive adherence' as a sort of unconditional assent. He stresses that belief in God is not a belief in an explanatory hypothesis, for that would involve a provisional and tentative adherence and this is 'alien to the whole spirit of religious belief. Having made our decision, we adhere to belief unconditionally, we commit ourselves as completely as one can ever commit oneself to anything.'[38] As with Lyas, the suggestion is that for the religious believer evidence is irrelevant to belief. This is not because to acknowledge the relevance of evidence would endanger the satisfaction of the motive from which one's religious interest is said to have arisen, but because, to cite Austin's statement of the argument, 'it would be religiously wrong, *faithless*, to abandon them in the face of contrary evidence, or even to contemplate the possibility of doing so.'[39] To be committed in this context is to be committed 'come what may'.

If this is true, then there is a sharp contrast with belief in a scientific hypothesis. For however attached the Newtonians were to their research programme, they did not have a commitment 'come what may'. If the world had started to behave differently, or if another theory were devised which exposed and coped with various anomalies in their own research programme, they would have felt obliged to give up their provisional commitment. And yet this is just what the person of religious faith is prohibited from doing, according to MacIntyre.

But is there really such a prohibition in the religious case? Does a commitment 'come what may' even make sense? Consider what such a commitment would demand of believers faced with evidence which seems to tell against their beliefs; say some scientific

---

[38] 'Logical Status of Religious Belief', 187.
[39] *Relevance of Natural Science to Theology*, 108.

discovery. Austin writes that

> it is not clear whether the MacIntyrean believer would (1) insist that the scientific discoveries are irrelevant, or (2) admit that logically they might be relevant but refuse to accept their implications. If he takes the former line, he will evidently need some further argument for the claim of irrelevance, beyond the bare fact that his faith forbids him to abandon any doctrine because of its apparent conflict with scientific assertions.[40]

Insisting that scientific discoveries are irrelevant to certain beliefs is proper enough. For example, in the exegesis of the creation stories in the book of Genesis, the apologist will rightly maintain that geological speculation about the age of the earth does not necessarily undermine the central assertion of the story, that we are dependent on God for our creation. But though this explanation of what is involved in believing in certain doctrines will relieve the tension with particular scientific discoveries in some cases, there is, of course, no guarantee that such explanations can always be provided, unless one maintains the thesis that no religious 'beliefs' are factual. In Chapter 4 it was shown that this thesis relies on the claim that resistance to evidence is a token that a belief is not factual. But it has been the burden of the last part of this chapter to suggest that it is a token of no such thing, for such resistance is characteristic of certain scientific hypotheses.

The second option is to say that though scientific discoveries are logically relevant to religious belief, believers refuse to accept the implications of such discoveries. That is, they accept that their beliefs may be contradicted by what is admitted to be good evidence, but claim that they do not feel compelled to come to some decision either in favour of the evidence or against the beliefs; they assert *p* at the same time as they acknowledge that *not-p* is to be believed too. Now there is no doubt that this may represent the state of mind of certain believers. Maybe they manage to 'believe' a contradiction by keeping their faith and their secular beliefs apart in two watertight compartments. But such intellectual schizophrenia is not characteristic of many believers, for evidence itself tells against the claim that they ignore the evidence. In fact they do take account of it and not just for apologetic purposes. To hold otherwise would be to ignore the history of theology's relations with science, the immense energy expended on devising a theodicy

[40] *Relevance of Natural Science to Theology*, 109.

and the development of critical study of the Bible. Each of these provides a stark example of a threat to belief causing acute intellectual difficulties, and so makes it surprising that it could be thought that the believer supposes that faith demands a confidence which involves turning a blind eye to the facts.

The assertion that the religious faith requires decisive adherence seems, then, to be false, and Mitchell has explained the notion of decisive adherence as stemming from a confusion. He notes that 'faith' has often been read as 'trust'. Hence to be faithless means to have failed someone by not placing in them the trust that they might expect. But whom have I failed if I come to doubt the existence of God? Mitchell writes:

although there is a Christian duty to trust in God, this does not imply a duty, let alone an unconditional duty, to go on believing that there is a God. Indeed, once it is admitted to be a genuine possibility that there is no God and that the case against his existence might become cumulatively overwhelming, it is pointless to maintain that one ought to go on believing nevertheless that there is a God, even when the belief could be seen to be false. It could not, in these circumstances be a duty owed to God, and there is no other conceivable reason why it should be a duty. This is to say that the requirement of unconditional faith is one which has its place within the system of theistic belief and cannot properly be interpreted as an obligation to continue to embrace the system itself.[41]

To put it in other words, one cannot have a duty to believe in God unless one admits his existence; it merely begs the question against the person who has doubts on this score to say that there is a duty to believe. One cannot admit the duty unless one believes that God exists, so one's duty cannot be to believe that God exists.

Note, however, that in the quotation just given, Mitchell, whilst arguing that a duty to believe in God (a duty of 'unconditional assent') 'cannot properly be interpreted as an obligation to embrace the system', none the less says that 'the requirement of unconditional faith is one which has its place within the system of theistic belief'. How are we to interpret this duty, in the light of our claim that religious belief may be founded on an explanatory justification?

There are two duties associated with living a life of faith which may give rise to the notion of unconditional faith. First, there is a

---

[41] *Justification of Religious Belief*, 139–40.

duty to trust in God, who having created and redeemed the world, loves and sustains it. Mitchell writes:

The faith that *is* unconditional is the believer's trust in God; it is this which forbids him to despair in the face of doubts and difficulties. This trust is the corollary of God's nature as he believes it to be. The God 'in whom is no variableness and shadow of turning' is a faithful and merciful God who will not abandon his creature.[42]

There is, secondly, another sort of unconditionality about faith. For a true faith will pursue wholeheartedly the obligations and actions appropriate to one who believes in God, without reserve or equivocation. In this sense faith or devotion to God might be described as unconditional, whereas devotion to one's country would not be.

Although we may describe faith as unconditional to capture the two aspects of it just mentioned, it is clear from the evidence concerning the nature of religious belief which we have reviewed that it is not unconditional in the sense MacIntyre implied, and indeed that the notion of a duty to believe is incoherent. At the prompting of Mitchell and Flew, MacIntyre has since repudiated his earlier views, saying that he 'came to see that position as importantly inconsistent with the very professions of that religious faith of which it professed to be the elucidation'.[43] In particular, since there is no duty to believe, the struggle with the evidence which is part of normal religious faith is not inconsistent with it. Thus, with the second premiss of MacIntyre's argument suitably qualified, the exponent of explanatory justification has no cause to reject it. For one can agree that religious faith has an element of unconditionality about it, but think that this is in no way inconsistent with acknowledging the possibility that argument or evidence may compel one to give up the assertion that God exists.

Conclusion

Phillips and Wittgenstein looked for support for their under-standing of religious discourse from the fact that religious belief seems to be endowed with a certain immunity to the evidence. If their expressive interpretation were the correct one, then of course

---

[42] 'Faith and Reason: A False Antithesis?', 141.
[43] *Metaphysical Beliefs*, ed. Toulmin *et al.*, Preface, p. xi.

the attempt to provide an explanatory justification of religious belief would be misguided and irrelevant. A problem for their understanding, however, is found to lie in the phenomenon of loss of belief, which seemed to support the traditional, intellectualist account. But if that latter account is to vindicate itself, it must try to reconcile the two poles of resistance and amenability to evidence.

The sense that religious belief is not conjectural also provides the grounds for a different attack on the religious adequacy of an explanatory justification. For if it is not a hypothesis, MacIntyre argues, then an explanatory justification would not suffice to ground religious belief. MacIntyre's evidence for the claim that religious belief is not a hypothesis lies primarily in an understanding of hypotheses in general as being tentative and provisional, standing over against his conception of religious belief as characteristically confident.

Wittgenstein, Phillips, and MacIntyre are answered by an elucidation first of the manner in which hypotheses in science may be held, and second of the type of commitment appropriate in the religious case. It has been argued that attention to both brings them closer together. Consideration of the high-level theories of science shows how mistaken is the notion that the scientist's commitment is provisional and tentative. And consideration of the demands of faith has shown that belief does not require 'unconditional assent' in the sense of undoubting certainty. Hence the secure status of religious belief in the believer's life need be explained *neither* by these beliefs not really being beliefs, *nor* by believers supposing that they possess reasons of greater weight for holding them than an explanatory justification could provide.

In a recent work on the philosophy of religion, which I cited in Chapter 1, we are told that:

Religious believers, unlike scientists, typically and characteristically seek to preserve their favoured models from criticism *at all costs* and *in the face of whatever difficulties* they encounter—something that would certainly be seen as irrational in a scientist.[44]

It is hoped that this chapter has revealed the simplistic character of such judgements, and has fostered a readiness to go forwards with the belief that an explanatory justification is at least not religiously

---

[44] O'Hear, *Experience, Explanation and Faith* (London: Routledge & Kegan Paul, 1984), 100.

inadequate, for it allows, in principle, for the confidence appropriate to religious belief. The problem of elucidating the conditions which an hypothesis must fulfil for that confidence to be warranted is the one to which I turn in Chapter 6.

6.

Theism and Inferring to the Best Explanation

CAN theism be defended—at least in principle—as an explanatory hypothesis? Chapters 4 and 5 have dealt with the charges that explanatory justification is either irrelevant to Christian theism or, if not strictly irrelevant, then religiously inadequate. These charges arise from a concern to protect the character of Christian belief, but there are doubts about an explanatory justification which have a different source. These doubts are suggested by the thought that any analogies there may be between the justification of theism and the defence of scientific theories are outweighed by the disanalogies. Here the suggestion is not that the proposed analogy depends upon a distortion of religious belief, but rather that it depends upon a distortion of the nature of science.

At this stage I am concerned not with detailed objections to a cumulative case for theism which argue, for example, that there are alternative explanations of each piece of evidence cited in favour of theism. Rather my concern is with objections to the notion that this justificatory route might, in principle, be successful. Two principal objections will be considered: the first, that theism could never qualify as an explanation. *A fortiori* it could never qualify as the best explanation. Second, even if it is in principle possible that theism should satisfy the conditions necessary to qualify as an explanation, it could not demonstrate that it is the best explanation and worthy of our belief. The nature of these objections will become clear as they are developed in sections 1 and 3 of this chapter. What is demanded in response to them is a consideration of the model of explanation presupposed in the first objection, and then attention to the account of inference to the best explanation presupposed by the second.

1. An Objection from the Nature of Explanation

Could theism constitute an explanation? One way in which to answer that question is to move from an informal characterization (such as 'explanations are answers to why-questions') towards a rigorous statement of a theory of explanation. This would specify requirements to be met for an answer to count as an explanation and against this theory, theistic 'explanation' could then be measured. The most important and influential proposals are those made by Hempel, proposals which, if they are read in a certain light, do indeed constitute a challenge to theism offering itself as an explanation.[1]

Hempel suggests what he calls the deductive–nomological model of explanation, the main outline of which is that an explanation of an event E is to be effected by a combination of:

(1) a set of statements of the initial conditions, C_1, C_2, ... C_n

(2) a set of general laws, L_1, L_2, ... L_n

which together allow for the deduction of the sentence to be explained, the explanandum, E. As an example of this scheme, Hempel considers the experiments carried out by Périer, who found that the length of the mercury column in a Torricelli barometer decreased with increasing altitude.[2] This explanandum Pascal explained by reference to the following claims, which 'have the character of general laws':

(1) at any location, the pressure that the mercury column in the closed branch of the Torricelli apparatus exerts upon the mercury equals the pressure exerted on the surface of the mercury in the open vessel by the column of air above it.

(2) the pressures exerted by the columns of mercury and of air are proportional to their weights; and the shorter the columns the smaller their weights.

Add to these general laws the important initial condition that as Périer carried the apparatus up the mountain the column of air above the open vessel became shorter, and it follows that the

---

[1] C. Hempel, 'Studies in the Logic of Explanation', in *Aspects of Scientific Explanation* (New York: Free Press, 1965), 245–95.

[2] C. Hempel, *Philosophy of Natural Science* (Englewood Cliffs, N J: Prentice-Hall, 1966), 49–50.

mercury in the column will decrease in height as there is an increase in altitude. To set it out as Hempel would:

$$L_1 \text{ and } L_2$$
$$C_1$$
$$\overline{}$$
$$E$$

So according to Hempel's simplest model, to explain is to show how the event in question may be deduced from initial conditions and laws. And by replacing 'may be deduced from' with 'is rendered highly probable by', Hempel gives essentially the same analysis of what he terms inductive-statistical explanation.

Having given the account I have set out above, Hempel comments that 'such arguments . . . constitute one type of scientific explanation.'[3] In earlier statements Hempel seems less cautious. Thus in the original piece, 'Studies in the Logic of Explanation', Hempel claims that: 'an explanation of a particular event is not fully adequate unless its explanans, if taken account of in time, could have served as a basis for predicting the event in question.'[4] The impression is given that the deductive-nomological model is not just descriptive of the form of some explanations, but that deductive-nomological explanations constitute the only legitimate form of explanation. All good explanation is conformable to this schema, so that conformity to the pattern set out above is both a necessary and a sufficient condition on explanation. In other words, as Hempel puts it, 'The decisive requirement for every sound explanation remains that it subsume the explanandum under general laws.'[5]

Debate about the adequacy of Hempel's account of explanation has been voluminous.[6] Fortunately it is not relevant to my purposes to consider the question as to whether the deductive-nomological model sets out sufficient conditions for an answer to qualify as an

---

[3] Ibid. 51.

[4] 'Studies in the Logic of Explanation', 249.

[5] Ibid. 258. In a Postscript to 'Studies in the Logic of Explanation' Hempel expresses surprise that he should have been read as claiming that the deductive-nomological model provides necessary conditions for explanation. But as the quotations above suggest, his early thoughts on explanation lack the qualifications in his later essay 'Aspects of Scientific Explanation', in *Aspects of Scientific Explanation*, 331–496.

[6] For a recent survey of the debate and a forceful argument that the model needs not reform but outright rejection, see W. Salmon, *Scientific Explanation and the Causal Structure of the World* (Princeton N J: Princeton University Press, 1984).

explanation, that is, whether all deductive-nomological 'explanations' are really explanatory.[7] Instead of worrying about that, we should ask whether 'the decisive requirement for every sound explanation' is 'that it subsume the explanandum under general laws.' For if it is true that conformity to the deductive-nomological model provides a necessary condition for explanation, then the charge that theism does not qualify as an explanation would be well placed. For theistic 'explanation' will be in terms of the intentions of a divine agent, and will not, therefore, have the form of explanation in terms of general laws.

Theistic explanation is a species of personal explanation, where we explain an event E as brought about intentionally by a rational agent P. Now it has been maintained that, in fact, personal explanation is reducible to the deductive-nomological pattern, but Richard Swinburne disputes this claim. Swinburne suggests that Davidson's 'is the most plausible form of reductionist theory', with its assertion that 'very roughly, actions are states which have occurrent mental states, desires for their occurrence, among their causes. Personal explanation is analysable in terms of the production of effects by such desires.'[8] Thus, supposing E is the result of a basic action:

Then to say that P brought E about intentionally is just to say that a state of P or an event involving P, P's intention that E occur, J, brought it about. To say that P had the power to bring E about is just to say that P's bodily condition Y (brain-states, muscle-states, etc.) and environmental conditions Z (no one having bound P's arm, etc.) and psycho-physiological laws L are such that an intention such as J is followed by the event intended, E.[9]

Hence it is alleged that personal explanation is, in essence, explanation according to Hempel's model.

Against this reductionist strategy, Swinburne argues that any analysis of intentional action which treats intentions (or wants, desires, or volitions) as the initial conditions which, under general laws, cause such and such an action, is unsuccessful, for the reason that

[7] For doubts about whether the deductive-nomological model provides sufficient conditions for explanation, see for example R. Harré, *The Principles of Scientific Thinking* (London: Macmillan, 1970), 16 f.

[8] *Existence of God*, 40.

[9] Ibid. 36.

an intention (or wish or desire) of *P* to bring about *E*, if it is some occurrent state or event, could bring about *E* without *P*'s having intentionally brought about *E*. Causation by an intention does not guarantee intentional action.[10]

For an example which makes the point Swinburne turns to Richard Taylor's objection to an analysis which takes a 'want' as the causal factor.

> Suppose . . . that a member of an audience keenly desires to attract the speaker's attention but, being shy, only fidgets uncomfortably in his seat and blushes. We may suppose, further, that he does attract the speaker's attention by his very fidgeting; but he did not fidget *in order* to catch the speaker's attention, even though he desired that result and might well have realized that such behaviour was going to produce it.[11]

The example shows that 'To say that *P* brought something about intentionally is not to say that some state of *P* or event involving *P*, such as an intention, brought that thing about.'[12] Thus an analysis of personal explanation according to this pattern leads to a failure to distinguish between an agent's bringing about something intentionally, and the agent's intention bringing something about.

Theistic explanation, as an instance of personal explanation, does not conform to the Hempelian model. Is it shown thereby to be inadmissible? Theists have usually resisted this inference, though they have often admitted, along with Mitchell, that 'theological explanations cannot be of a scientific kind.'[13] The reasons, however, which might be used to justify resistance to the Hempelian model, would also show that Mitchell's concession is misplaced. For theists will insist on the integrity of their own explanations by suggesting that the failure of religious explanations to conform to the deductive-nomological model is a failure they share with many scientific explanations. Hence in insisting on the integrity of the explanations they employ, it becomes apparent that theists should not endorse the contrast between 'religious' and 'scientific', or 'personal' and 'scientific' explanations (where, for 'scientific' read 'deductive-nomological'), which both Mitchell and Swinburne countenance.

---

[10] Ibid. 40.

[11] R. Taylor, *Action and Purpose* (Englewood Cliffs, N J: Prentice-Hall, 1966), 248; cited by Swinburne, *Existence of God*, 40.

[12] *Existence of God*, 41.

[13] *Justification of Religious Belief*, 101.

The point is that it is not just personal explanations which fail to conform to Hempel's type, but also many explanations given in science.[14] Harré cites evolutionary biology as one example which fits ill with the approved pattern.[15] And he is no doubt right. The fact that marsupials are presently found exclusively in Australia is explained in evolutionary theory by a long story of competition, geographical isolation, and adaptation. Although we are inclined to accept the account as an explanation it does not in the least approximate to a deduction, mainly because the initial conditions are not known in enough detail.[16] Unless we are to dismiss this as pseudo-explanation, we have to reject Hempel's schema as a necessary condition on explanation.

Hempel would reply that the evolutionary account is an 'explanation sketch'; he was not guilty of supposing that all explanations in the sciences are demonstrably deductive-nomological. His claim was that though not presently displayable according to the deductive-nomological form, acceptable explanations are, in principle, so expressible. But this concession seems to dissolve any prescriptive force which Hempel's model might possess in virtue of its representing the type of explanation characteristic of the natural sciences. For what does this 'in principle' amount to? Short of giving a deductive-nomological account of an event it remains questionable that an explanation can be given in this form. Hence it can hardly be by reference to the 'in principle' possibility of an explanation conforming to the model that we acknowledge it to be an explanation. Our recognition of it as such must, in fact, be quite independent of the model.

This should give comfort to the theist, who will concede, as we have said, that theism aims to give an explanation which is not, and cannot be, expressed in deductive-nomological form. The inadequacy of Hempel's model may be established well enough by attention to the nature of personal explanation and insistence on its integrity. But the point is stronger when it is stressed that our understanding of something as an explanation is independent of its conformity to, or 'in principle conformity to', the deductive-

---

[14] This is argued at length by E. L. Schoen in *Religious Explanations: A Model from the Sciences* (Durham, N C: Duke University Press, 1985).

[15] *Principles of Scientific Thinking*, 19.

[16] See P. Kitcher, *Abusing Science* (Milton Keynes: Open University Press, 1983), 140.

nomological model. And this can be shown by reference to science as much as by reference to religious or common-sense explanations.

Hempel's account of explanation seemed to demand of explanation that it be in terms of deduction from general laws. It is true that theistic explanation will not be of this sort, but also that our recognition of something as an explanation is independent of its conformity to the deductive-nomological model. What that recognition consists in I do not hazard to say, for I have no general theory of explanation to offer. I rest content with the claim that theism recognizably has the form of an explanation, whether or not we deem it an explanation supported by sufficient evidence to render it acceptable.

2. Darwin and Inferring to the Best Explanation

The first objection held that theism was such that it could not, in principle, be characterized as an explanation—*a fortiori* it could not be justified as an inference to the best explanation. The second objection comes at things from the other way. Here the point is that even if theism is an explanation of sorts, its character is such that it could never establish that it is an explanation worthy of belief. In particular a charge against theism may be based on Lakatos's demand that to be worthy of belief explanations must yield confirmed predictions, and in section 3 of this chapter this demand will be discussed.

This second type of objection, however, raises a larger question. How does an explanation establish itself as worthy of belief? What makes it acceptable? The correctness of some explanations seems entirely obvious. The paper burnt because a match was put to it. The car started because the key was turned. But in a number of cases our believing an explanation to be true consciously depends on the sort of inference which I have been following Gilbert Harman in calling 'inference to the best explanation'.

Stating the conditions under which such an inference is valid and establishes the belief-worthiness of an explanation is a complex task and I shall not attempt to give a general, abstract answer to the problem. Rather, I shall pay attention to those suggested answers which seem to have particular relevance to theism itself, ones which prima facie raise doubts about whether theism could be given an explanatory justification. A critical account of inference to the best explanation will, then, emerge only negatively.

In considering accounts of explanatory justification it will prove useful to have at our disposal an example of a purported inference to the best explanation and for this purpose I shall expound briefly the argument of Darwin's *Origin of Species*. In his conclusion to *The Origin of Species* Darwin says that 'this whole volume is one long argument'.[17] And it is indeed a brilliant piece of sustained and powerful persuasion, which will not only serve in section 3 of this chapter as a model against which theoretical accounts of explanatory justification may be compared, but will be relevant also in a subsequent chapter when certain substantial objections to an inference to theism will be mooted.

'Light has been thrown on several facts, which on the theory of independent creation are utterly obscure'—thus Darwin on the early chapters of *The Origin of Species*.[18] It is one of the most engaging aspects of this work, that Darwin is studiously concerned not to overstate his case; to the outsider it might seem more accurate to say that Darwin has heaped up observations which, taken together, put the theory of special creation in a hopeless position.

These observations are so numerous that I shall give as examples only two classes of them. The first has to do with the fact that it is possible, and indeed seems natural, to divide organic beings according to principles of what we might term 'family resemblance'. Moreover, Darwin notes that subordinate groups in a class cannot be ranked in single file, but seem rather to be clustered round points, and these round other points, and so on. He comments that 'On the view that each species has been independently created, I can see no explanation of this great fact in the classification of all organic beings.'[19] The clustering which Darwin observes leads him to the conclusion that

nature is prodigal in variety, but niggard in innovation. Why, on the theory of Creation, should this be so? Why should all parts and organs of many independent beings, each supposed to have been separately created for its proper place in nature, be so invariably linked together by graduated steps?

[17] C. Darwin, *The Origin of Species* (Harmondsworth: Penguin, 1968; 1st pub. 1859), 435.
[18] Ibid. 230.
[19] Ibid. 171.

Why should not Nature have taken a leap from structure to structure? On the theory of natural selection we can clearly understand why she should not; for natural selection can act only by taking advantage of slight successive variations; she can never take a leap but must advance by the shortest route and slowest steps.[20]

So those who believe in independent acts of creation must face the question posed by Darwin:

What can be more curious than that the hand of man, formed for grasping, that of a mole for digging, the leg of the horse, the paddle of the porpoise, and the wing of the bat, should all be constructed on the same pattern, and should include the same bones, in the same relative positions?[21]

Similar morphological questions appear throughout the work. Why does each species of the equine genus tend to stripes?[22] Why should the brain be enclosed in a box composed of 'such numerous and extraordinarily shaped pieces of bone'?[23] And why should a crustacean 'which has an extremely complex mouth formed of many parts, consequently always have fewer legs; or conversely, those with many legs have simpler mouths'?[24]

  Apart from facts of 'family resemblance', a second great class of observations has to do with the distribution of species. A central problem for the theory of special creation is to explain why similarity of climate and conditions does not imply similarity of flora and fauna.[25] To take a specific example again, Darwin is concerned by the absence of frogs, toads, and newts from so many oceanic islands which are otherwise suitable for their flourishing— witness how well such creatures have done when introduced, for example, to Madeira. 'Why, on the theory of creation, they should not have been created there, it would be very difficult to explain.'[26]

  As is well known, Darwin was much impressed by the distinctive (though American) flora and fauna of the Galapagos Islands, so utterly dissimilar from, say, the (African) inhabitants of the climatically similar Cape Verde Islands. But it was not only the fact of geographical similarity and zoological dissimilarity which impressed him. Equally he was impressed by facts of relative similarity between many island dwellers and their mainland

[20] Ibid. 223–4.   [24] Ibid. 418.
[21] Ibid. 415.   [25] Ibid. 365.
[22] Ibid. 201.   [26] Ibid. 382.
[23] Ibid. 417.

neighbours and he even claimed to observe a correlation between the depth of sea between islands and shore, and the resemblance between the islands' mammals and those of the mainland.[27] As Darwin never tires of reminding us, such facts and relationships seem inexplicable on the view of independent acts of creation.

I have drawn attention, then, to two classes of facts which Darwin makes much of. Morphological studies reveal gradations of structural resemblance between creatures (some now extinct); study of geographical patterns of distribution suggests relationships of a complexity which cannot be accounted for merely by reference to climate and conditions. The point of these examples is at one and the same time to draw attention to the limitations of the theory of independent creation, and to raise the stock of Darwin's own theory of evolution by natural selection. For according to Darwin, the theory of evolution by natural selection offers explanations where previously there was inscrutability.

Though Darwin's theory is sometimes referred to as 'the theory of evolution', this label fails to note that which makes Darwin's work distinctive, namely, its suggestion of a mechanism whereby the long-known fact of evolution is rendered intelligible. That mechanism is natural selection, and Darwin's account of it is modelled on domestic selection. In domestic selection breeders choose which of a species' variable features to foster by controlling the breeding of their stock. In nature it is the shortage of resources or the 'severe struggle for life' which controls the breeding stock, and by this control ensures that those best suited to survive (that is, those exemplifying advantageous variations) have a greater chance of leaving survivors. Hence 'by the strong principle of inheritance they will tend to produce offspring similarly characterised', with a resulting modification and adaption of the natural stock.[28]

It is by reference to this model, so Darwin alleges, that the observations which discredit the theory of independent acts of creation are to be understood. Both the morphological and the geographical relationships of organic beings are intelligible on the theory of natural selection. Darwin acknowledges with great candour the problems which are implicit in his account. He says that 'a crowd of difficulties will have occurred to the reader. Some

---

[27] *The Origin of Species*, 384.
[28] Ibid. *passim*, but esp. chs. 1–4.

of them are so grave that to this day I can never reflect on them without being staggered; but', he continues, 'to the best of my judgement, the greater number are only apparent, and those that are real are not, I think, fatal to my theory.'[29] And as large a part of *The Origin of Species* is devoted to dealing with difficulties for the theory of evolution as to spelling out the inadequacies of the theory of special creation. In chapters 6 and 7 Darwin takes note of 'the extreme perfection' of individual organs such as the eye, and of particular instincts such as that of bees in constructing a cell, and tries to persuade us of the possibility of their gradual formation through the preservation of favourable variations. That possibility presupposes the existence of numerous intermediate forms and in a chapter entitled 'On the Imperfection of the Geological Record', Darwin attempts to account for the relative paucity of the fossil remains which leave little evidence of transitional species. No opponent of Darwin could have put the latter objection with greater force:

During each of these years, over the whole world, the land and the water has been peopled by hosts of living forms. What an infinite number of generations, which the mind cannot grasp, must have succeeded each other in the long roll of years! Now turn to our richest geological museums, and what a paltry display we behold![30]

The two objections to which I have made reference are related. An inability to conceive of the slow evolution of the eye, for example, would be overcome by evidence of intermediaries which would witness to that evolution. To account for the lack of such evidence, Darwin takes up a metaphor from Lyell, and refers to the natural geological record as 'a history of the world imperfectly kept, and written in a changing dialect; of this history we possess the last volume alone . . . Of this volume, only here and there a short chapter has been preserved; and of each page, only here and there a few lines.'[31] So concerning the reader who still has doubts about the formation of perfection through natural selection, Darwin counsels that 'His reason ought to conquer his imagination; though [a characteristic qualification] I have felt the difficulty far too keenly to be surprised at any degree of hesitation in

---

[29] Ibid. 205.
[30] Ibid. 297.
[31] Ibid. 316.

extending the principle of natural selection to such startling lengths.'[32]

Darwin's *Origin of Species* is a wonderful example of a purported inference to the best explanation, the force of which cannot fully be appreciated by an abstract of only some of the points he makes. Surveying a wide field of facts concerning natural history, Darwin looks for the best explanation of them, argues that special creation fails at many points, and defends the theory of natural selection as to be preferred.

But is it possible to give a theoretical account of what it is for an explanation to be the best explanation, the explanation which is more probably true than any other? I have noted in a previous chapter that whilst Harman draws our attention to inference to the best explanation as a commonplace and valid form of reasoning, he doubts that the conditions for its proper employment are at all obvious. He writes:

There is, of course, a problem about how one is to judge that one hypothesis is sufficiently better than another hypothesis. Presumably such a judgement will be based on considerations such as which hypothesis is simpler, which is more plausible, which explains more, which is less *ad hoc*, and so forth. I do not wish to deny that there is a problem about explaining the exact nature of these considerations; I will not, however, say anything more about this problem.[33]

Unlike Harman we must say something more about the considerations which are thought to count for and against hypotheses. In particular I shall concentrate upon a criterion from Harman's list— absence of *ad hoc*-ness—which is especially relevant to the possibility of an explanatory justification of theism, and which seems to speak against that possibility. In determining what force is to be given to the consideration that one is to prefer a theory to a rival in so far as it is less *ad hoc*, we shall also elucidate the demands of explanatory power and plausibility. Darwin's *Origin of Species* will provide a useful example against which to weigh the suggestions we encounter.

---

[32] *The Origin of Species*, 219.
[33] 'Inference to the Best Explanation', 89.

3. Explanatory Power and an Objection to Ad Hoc-ness

Harman suggests that a theory is to be preferred to its rivals according to whether it is plausible, more explanatory, less *ad hoc*, or simpler. I note a significant omission from this list—conceptual coherence or consistency—only to put it on one side. This is not because the criterion of coherence has been unimportant in scientific discussions regarding choice between theories, nor because it is unimportant for our specific concern with the propriety of belief in theism. The question of the coherence of the concept of God is an age-old subject for debate which has found such modern protagonists as Kenny and Swinburne.[34] But reasons for not delaying over this consideration, besides the pragmatic reason provided by the very scope of the issue, are as follows. On the one hand the problem of consistency is not one which arises especially for theism as against scientific theories. And on the other, it is only if one were sure that every possible concept of God to be found within the tradition of Christian theism is incoherent that one could ignore the case for his existence. Thus J. L. Mackie in *The Miracle of Theism* engages directly with the arguments in favour of God's existence, whilst maintaining that there are serious conceptual problems with theistic belief. One such—that of reconciling the attributes of omniscience, omnipotence, and goodness traditionally predicated of God, with the existence of evil—will be discussed in the next chapter. But clearly Mackie does not regard the conceptual problems to be of such a sort as to defeat theism at the outset.

The claim that theism is indefensible as an inference to the best explanation cannot, then, without a good deal of argument, rely on the incoherence of the concept of God. But there is a criterion of theory choice in Harman's list which might be thought to threaten theism more obviously. That is absence of *ad hoc*-ness. How does this consideration enter into the choice between theories if, that is, it does so enter? Answering that question requires that we pay some attention to explanatory power, since a rule against *ad hoc*-ness is related to it as a check or qualification.

A theory's explanatory power is measured by its observational

[34] Kenny, *God of the Philosophers*, and R. Swinburne, *The Coherence of Theism* (Oxford: Clarendon Press, 1977).

success. Observational success has two components, retrospective and predictive:

    (i) success in accounting for known observations:

    (ii) success in suggesting new and corroborated observations.

A theory has the greater explanatory power the wider is the range of its ability to account for phenomena previously unconnected and to suggest and account for unexpected phenomena.[35]

A theory is *ad hoc* (that is, 'to this specific purpose'), if it is a mere response to observations which have already been made; that is, if it accounts only for previously known phenomena. Contrariwise, it is not *ad hoc* in relation to observations which it not only explains but also predicts. Thus Newton's theory was *ad hoc* in relation to the known movements of the planets, but not in relation to, say, the movements of newly observed comets.

It follows that any rule against *ad hoc* theories represents a qualification of the notion that explanatory power counts in a theory's favour. It implies that it is not explanatory power in a general sense which adds to a theory's credit. Rather, credit accrues to a theory only if it has success of the second type, namely, predictive success; otherwise it is *ad hoc*. In other words success of type (ii) is necessary for acceptance of a theory. (Success of type (i) is of course, uncontroversially regarded as a necessary condition of such acceptance.)

In Lakatos's philosophy success of the second type becomes pre-eminent, for it is held that it is only the generation and corroboration of novel predictions which adds to a theory's credit. He writes:

A research programme is either progressive or degenerating. It is *theoretically progressive* if each modification leads to new unexpected predictions and it is *empirically progressive* if at least some of these novel predictions are corroborated. It is always easy for a scientist to deal with a *given* anomaly by making suitable adjustments to his programme (e.g. by adding an epicycle). Such manoeuvres are *ad hoc*, and the programme is degenerating, unless they not only explain the given facts they were intended to explain but also predict some new facts as well.[36]

[35] This statement would be too vague for the advocate of the deductive-nomological model. According to this model explanatory success consists in the deduction of the explanandum from the explanans, and in the deduction of new and corroborated predictions. This tighter model of explanation would therefore reject the formulation which speaks merely of 'accounting for' and 'suggesting observations'.

[36] I. Lakatos and E. Zahar, 'Why Did Copernicus's Research Programme

A good example of a theory progressive in both senses and so fulfilling Lakatos's demand for corroboration by novel prediction is Fresnel's wave theory of light. Fresnel wrote an essay on the nature of refraction which was read by Poisson, a convinced corpuscularian, who noted that if Fresnel's theory were correct then the centre of a shadow of a disc would be marked by a bright spot. As Laudan notes of this celebrated case:

This predicted result was highly unlikely; it contradicted both the corpuscular theory *and* the scientist's intuitive sense of what was 'natural'. Indeed, the fact that the wave theory possessed this bizarre consequence was seen, *prior* to performing the experiment, as a kind of *reductio ad absurdum* of it. But when the appropriate tests were performed, the wave theory was vindicated by a concordance between what it predicted and the observed results.[37]

This example is taken from a paper in which Laudan discusses the eighteenth- and nineteenth-century debates about 'imperceptible fluids' and the epistemological questions which these debates raised, focusing 'chiefly upon the wave theory of light (with its seemingly attendant commitment to a luminiferous ether)':

On the epistemological side, most of the interest centers around the emergence of a new methodological criterion for evaluating hypotheses. In brief, this criterion, which was nowhere prominent in the late 18th-century debates about the methodological credentials of subtle fluids, amounts to the claim that an hypothesis which successfully predicts future states of affairs (particularly if those states are 'surprising' ones), or which explains phenomena it was not specifically designed to explain, acquires thereby a legitimacy which hypotheses which merely explain what is already known generally do not possess. The major figures in this story are Herschel, Whewell and Mill.[38]

If we take seriously the rule against *ad hoc* theories propounded by Lakatos and his predecessors, we say that it is only by the

Supersede Ptolemy's?', in I. Lakatos, *Philosophical Papers, i. The Methodology of Scientific Research Programmes*, ed. J. Worrall and G. Currie (Cambridge: Cambridge University Press, 1978), 179. Similar demands of legitimate theories are made by Popper in his 'Three Views Concerning Human Knowledge', in *Conjectures and Refutations*, 97–119.

[37] L. Laudan, 'The Epistemology of Light: Some Methodological Issues in the Subtle Fluids Debate', in *Science and Hypothesis* (Dordrecht: Reidel, 1981), 128.

[38] Ibid. 127. Historical precedent is thus clearly established for the solution to the problem of *ad hoc* theories and confirmation more recently propounded by Lakatos and Popper. See also Laudan's 'William Whewell on the Consilience of Inductions', in *Science and Hypothesis*, 163–91.

generation of novel and confirmed predictions that a theory gains credit. But if, following Lakatos, we make this a criterion of an acceptable theory and so rid ourselves of belief in *ad hoc* explanations, we have adopted a standard which will be epistem- ologically parsimonious. It is certainly a standard which, to put it at the very least, theism may find it difficult to meet. Whilst it may be that theism yields 'predictions' of a sort—that there is a life beyond this one, that our sins will be forgiven, and so on—at first sight these are either predictions for which there is no agreed means of confirmation, or predictions the falsification of which may well lie in the distant future, leaving us meanwhile without grounds for belief. If theism has to provide corroborated predictions before it merits our adherence then the case for theism is that much more difficult to make. Theism is manifestly *ad hoc*, as Hume pointed out long ago, prompting the conclusion that even if theism is explanatory, it could not show that it is worthy of belief.[39]

For Lakatos the reason for instituting this severe test for theories to overcome lies in his contention that it solves what he regards as 'the central problem in philosophy of science', namely 'the problem of normative appraisal of scientific theories; and in particular, the problem of stating *universal* conditions under which a theory is scientific'.[40] The problem of appraisal turns on Lakatos's accept- ance of what he terms 'conventionalism', better known as the Duhem-Quine thesis that any theory may be saved against refutation as long as one makes the necessary adjustments in one's beliefs to maintain consistency.[41] If a theory can be devised to account for any amount of data given enough ingenuity, then it seems to Lakatos that fit with evidence is not especially to the theory's credit. Its success may be *ad hoc*. A theory escapes condemnation by Lakatos only if it yields novel, confirmed predictions.

[39] For Hume's comments see his *Enquiry*, 146. More recently A. O'Hear bases his attack on the notion that religious explanations could be like scientific ones on the same point. See, for example, his remark: 'The real problem with the religious interpretation of religious experience is that it is quite deficient in explanatory power. The judgement that one has had a divine experience is quite unlike the judgement that one has seen a table in that it appears to lead to no testable independent predictions.' *Experience, Explanation and Faith*, 44. It will have occurred to the reader that the case for rational realism outlined in Chapter 3 is *ad hoc* too.

[40] 'Why Did Copernicus's Research Programme Supersede Ptolemy's?', 168.

[41] Ibid. 180.

In fact Lakatos's manœuvre fails to solve the very problem it was designed to solve, for just as any number of theories may fit with known observations, so any number of theories may yield confirmed predictions. The answer lies in the realization that a theory is not confirmed by its instances alone, a point which will emerge in the discussion of plausibility and consilience below.

That the rule against *ad hoc* theories also serves to exclude as unscientific theories such as theism, reveals its role in solving the 'problem of stating universal conditions under which a theory is scientific', which for Lakatos and Popper means acceptable.[42] Indeed, the very reason why the rule is so vigorously advocated by Popper and his one-time disciple Lakatos has to do with their shared preoccupation with devising a line of demarcation between science and non-science. Both are concerned with the pretensions of Marxism and Freudianism to the status of science. Their difficulties were in devising a criterion strict enough to rule out these two theories and yet flexible enough to allow in those theories which the scientific community regarded as acceptable. With the rule against *ad hoc* theories Popper and Lakatos thought they had found it.

Theories such as theism, which we may be tempted to accept on the basis of an inference to the best explanation, fail to satisfy this criterion. In the nature of the case, their observational success is retrospective. But if our account of the form of Darwin's *Origin of Species* is correct, then this classical example of scientific argument is also to be regarded as 'unacceptable' or inadequate according to Popper and Lakatos. Darwin's theory was *ad hoc*, for the case he made for it, as we have seen, rests largely on its superior explanatory power in relation to known facts.

This engenders the suspicion that Lakatos's standard is too severe, and that it will involve him in some difficulties in accounting for scientific behaviour in relation to the acceptance of theories. And this indeed proves to be the case. Take, for example, the adoption of Einstein's theories, where Lakatos naturally wants to hold that the replacement of Newtonian Gravitational Theory by

---

[42] For this conflation of the scientific and the acceptable see, for example, 'Falsification and the Methodology of Scientific Research Programmes', 116: 'For the sophisticated falsificationist a theory is "acceptable" or "scientific" only if . . .'.

the General Theory was, in his terminology, 'a progressive problem shift'. He writes that it is such

because it [general relativity] explained everything that Newton's theory had successfully explained, and it explained also *to some extent* some known anomalies and, in addition, forbade events like transmission of light along straight lines near large masses about which Newton's theory had said nothing but which had been permitted by other well-corroborated scientific theories of the day; moreover, *at least some* of the unexpected excess Einsteinian content was in fact *corroborated* (for instance, by the eclipse experiments).[43]

For Lakatos the 'moreover' which begins the last clause is crucial. It is the 'corroborated excess' of Einstein's theory which was necessary for its acceptance to be legitimate. And yet, as Glymour comments, 'This is not a wholly accurate capsule history.' For one thing the General Theory was barely theoretically progressive: 'Laplace was able to demonstrate the possibility of gravitational bending of light using Newtonian theory, and J. Soldner had calculated the value of the deflection by a massive body; only neglect made Einstein's prediction seem so very novel.' And it was not until very recently that the General Theory was shown to be empirically progressive:

The history of the eclipse measurements, like the history of the measurements of the gravitational red shift of sunlight, is one of uncertainty and obscurity. A 1918 eclipse observation obtained results contradicting Einstein. Of the two 1919 eclipse expeditions, one obtained results in at best fair accord with Einstein's theory and these results certainly had a great influence on contemporary opinion; an American eclipse expedition in 1922 also obtained results close to, but significantly higher than those predicted by general relativity; some later measurements went against Einstein. The preponderance of observations of the red shift of sunlight went against Einstein. Until the gravitational red-shift measurements of the 1960s using the then newly discovered Mössbauer effect, the chief empirical grounds for general relativity were that it could explain two phenomena which it certainly had not predicted: the anomalous advance of the perihelion of Mercury and the cosmological red shifts discovered by Hubble. The latter, at least, was a novel phenomenon.[44]

Contrary then to what Lakatos implies, Einstein's predictions were

---

[43] 'Falsification and the Methodology of Scientific Research Programmes', 124.

[44] C. Glymour, *Theory and Evidence* (Princeton, N J: Princeton University Press, 1980), 98–9.

neither novel nor confirmed prior to the 1960s. And yet his theory was accepted long before the measurements of the 1960s which according to Lakatos represent the vital test of the theory. The theory got on well without any confirmed predictions, its ability to solve old problems being quite enough.

The same point is made by Putnam in relation to the acceptance of Newton's theory of universal gravitation. The general acclamation of the theory stemmed from Newton's ability to derive Kepler's laws, account for the tides on the basis of the gravitational pull of the moon, and cope with small perturbations in the orbits of the planets, none of which constituted a corroboration in the sense intended by Popper or Lakatos. We have to wait for some hundred years after the theory was introduced and accepted for anything which resembles a Popperian test.[45]

The points made by Glymour were forced upon Lakatos by Zahar's study of the support given to Einstein's theory by its explanation of the long outstanding anomaly of Mercury's perihelion.[46] In response to this study Lakatos added to his rule against *ad hoc*-ness the qualification that success in accounting for a known phenomenon *did* count in a theory's favour, so long as the theory was not designed to cope with the fact it now explains and which is subsequently cited as part of its explanatory power.[47]

According to Lakatos's revision, the true test of a theory's explanatory power is in the generation of—what are for that theory at least—new (that is, unpremeditated) and confirmed predictions. Lakatos now allows then, that retrospective observational success undoubtedly and quite properly leads to a theory's acceptance, so long as the theory was not designed with these known facts in mind.

But this concession, if it accommodates the case of Einstein, accounts for neither Newton nor Darwin. Short of some detailed psychological investigations we cannot say which facts Darwin intended to explain and which he merely happened to explain unintentionally. And since appraisal of the theory allegedly hangs on these intentions we are at a loss in assessing it. Our suspicion may well be that, as with theism, Darwin's theory is designed with

---

[45] 'The "Corroboration" of Theories', 267.
[46] E. Zahar, 'Why Did Einstein's Programme Supersede Lorentz's?', *British Journal for the Philosophy of Science*, 24 (1973), 95–123 and 223–62.
[47] 'Why Did Copernicus's Research Programme Supersede Ptolemy's?', 185.

the facts in mind, as a response to them. But to settle whether or not the theory is acceptable we would need, according to Lakatos's proposal, to engage in painstaking historical research.

Yet that only serves to show that even this revised test proves to be of unjustified severity. It is not a *natural* response on reading Darwin's *Origin of Species* to determine temporal relations between the theory's conception and Darwin's knowledge of the facts he claims to explain, nor is it a *proper* response. The acceptability of a theory can hardly depend on the intentions of its designer. If it were true that it did, the consequence would be that the same theory (independently formulated by two scientists one of whom had the intention to explain a more limited range of facts than the other), would be acceptable as proposed by one, unacceptable when proposed by the other.

We have been asking whether success in suggesting new and corroborated predictions is a necessary condition for the acceptance of a theory (where, in the light of Lakatos's revision, 'new' may be taken weakly as meaning 'not foreseen in the design of the theory'). If it is a necessary condition, then belief in theism based on an inference to the best explanation would be rendered improper— unless we can find a theist who proposed theism and only then discovered that it explains the classical evidence for theism! Our argument suggests, however, that it is not a necessary condition. But that leaves us with another problem. If we reject Lakatos's necessary condition for theory acceptance are we maintaining that success in accounting for known observations is itself sufficient? For this, it seems, is too generous. If it allows theism as in principle acceptable as an explanation, it does so by allowing in too much else besides.

This worry about the nature of success brings us back to a problem alluded to in Chapter 3, namely the apparent vagueness of 'success' in the realist's vocabulary. To quell these doubts the realist adds that there are two checks on acceptance of a theory in virtue of its explanatory power. First of all a theory's explanatory power must have a certain range; and second, a theory must not only have sufficient explanatory range, but must also be judged plausible.

Newton-Smith suggests that presumably Lakatos adds his proviso about the theorist's intentions because 'he does not wish to give positive appraisal to a theory designed *ad hoc* to account for some single known fact.' But, he continues:

this can be precluded without having to make our theory appraisal dependent on an evaluation of the intention of the theoretician. For if we are dealing with a theory which depends for its value more on the explanation of known facts than on the corroborated prediction of novel facts, its positive worth will depend on the range and diversity of the facts that it accounts for. Consequently, a theory cooked up to 'explain' a single fact will not be meritorious. If, however, it gives a unified explanation of a diverse range of facts not previously known to be connected, the theory will have merit whether or not the scientist's intention was simply to have a theory which would do just this.[48]

In this passage Newton-Smith recognizes that, as was mentioned in Chapter 3, the realist implies more by 'success' than just accounting for known observations—something more like what Whewell drew attention to with his notion of 'the consilience of inductions'.

By 'induction' Whewell described what Peirce meant by abduction, and Harman by inference to the best explanation. Whewell writes: '*The Consilience of Inductions* takes place when an Induction, obtained from one class of facts, coincides with an Induction, obtained from another different class. This Consilience is a test of the truth of a Theory in which it occurs.'[49] The idea expressed here seems to be that a theory is acceptable in so far as it is supported by convergent lines of argument. So, in Laudan's paraphrase, a consilience of inductions occurs when 'two chains of "inductive reasoning" from seemingly different classes of phenomena lead us to the same "conclusion"'.[50] Elsewhere Whewell likens the consilience of inductions to 'the testimony of two witnesses in behalf of the hypothesis; and in proportion as these two witnesses are separate and independent, the conviction produced by their agreement is more and more complete'.[51]

Whilst the point is intuitively plain, there is of course a problem

---

[48] *Rationality of Science*, 88.

[49] W. Whewell, *The Philosophy of the Inductive Sciences, Founded upon Their History*, 2nd edn. (London: Parker, 1847), ii, 469. Cited by Laudan in 'William Whewell on the Consilience of Inductions', 164.

[50] Laudan, 'William Whewell on the Consilience of Inductions', 164. But see the following pages where Laudan elaborates further what Whewell included under the heading 'the consilience of inductions' and produces stricter tests of the sort favoured by Lakatos, such as that a theory should predict cases of a type unthought of in the formation of the hypothesis. Obviously I do not endorse this aspect of Whewell's treatment of consilience if this is regarded as a necessary condition for it.

[51] Whewell, *The Philosophy of the Inductive Sciences*, ii, 285; cited by Laudan, 'William Whewell on the Consilience of Inductions', 170.

with giving a more technical account of what is meant by 'different classes of phenomena', and of their being 'separate and independent'. We may readily grant that in relation to Darwin's theory the morphological facts and the facts of geographical distribution are two classes of fact which point in the same direction. But other cases may not be so sharp. Do facts about the bones in a mole's hands and those in a dolphin's fin constitute 'separate and independent' classes of fact, or are they rather to be described as belonging to a single class of facts concerning mammalian bone-formation? In any given case the distinction may be crucial and would demand a more adequate criterion for distinguishing classes of facts.

Though I have given considerable attention to the role of explanatory power in the choice of explanations, it does not stand alone. Explanatory power is controlled by plausibility, which functions as a check on theories and serves as a basis for preferring one to another. Thus we might prefer a theory with a lesser range of explanatory power because of its plausibility and fit with background knowledge.

In the case of the theory of evolution by natural selection, its plausibility depends on assumptions about, for example, the age of the earth. Evolution by the means suggested by Darwin's model requires there to have been countless generations of species modified by the action of environment on random variation; hence Lord Kelvin's estimate of the age of the earth (based on calculations of the supposed rate of cooling of the earth's crust) which was significantly lower than Lyell's on which Darwin relied, called into question the basis of the theory. Darwin's theory declined in plausibility as Lord Kelvin's star rose.[52] And it is by reference to such arguments that latter-day creationists seek to establish their own position as 'scientific'.[53] For if the age of the earth were demonstrated to be significantly less than Darwin requires, then no matter the potential explanatory power, the theory would be unacceptable.

Another aspect of the plausibility of Darwin's theory of evolution by natural selection turns on the possibility of nature evolving through adaptation to environment. Darwin's bid to convince us of

---

[52] See J. W. Burrow, 'Editor's Introduction' in *The Origin of Species*, 46–7.
[53] See Kitcher, *Abusing Science*, 100–1.

the plausibility of this suggestion depends on pressing as far as possible the analogy between the undoubted facts of domestic selection and his own conception of the workings of nature to the same effect.

To suppose, however, that explanatory power and plausibility always stand in a straightforward and simple relationship is to make the mistake of naïve falsificationism which was discussed in the previous chapter. For judgements of plausibility will make appeal to background *theories* which themselves rest on an inference to the best explanation and can be challenged and called in question by another theory with high explanatory power. Thus Darwinians might have said that the explanatory power of their theory was such that it was Kelvin's which was imperilled by the lack of fit, not their own. And history would have shown them to be right. Thus though the credit a theory derives from its explanatory power is usually related to its plausibility, judgements of its plausibility may be crucially influenced by estimates of explanatory power.

Nevertheless, plausibility enters in as a factor in the choice of a putative explanation as acceptable, and may often be a decisive factor where explanatory power is not enough to settle the issue. Imagine that we are confronted by two theories both of which seem to explain the same facts. Or imagine that there are two theories which are incompatible, but have their successes in different fields; or again, that one theory explains or promises to explain more, but yields less accurate predictions than another; and so on. In deciding between such competing theories, plausibility may be the key consideration.

So much for the first of Harman's criteria which determine whether or not an inference to an explanation counts as an inference to a probably true explanation. The requirement that a theory possess explanatory power is a demand for observational success. Rejecting the stipulation against *ad hoc* theories, the realist makes the requirement more taxing by the provision that observational success be significant. Lakatos is right, then, in that explanatory power is not an unqualified guide to acceptable theories, but his attempt to strengthen this insight into a prohibition of *ad hoc* hypotheses is seen to be unwarranted. The fact of a theory's being *ad hoc* is not in itself sufficient to rule out acceptance of that theory and the explanation it offers as probably true, though the propriety of that acceptance depends on the explanation being plausible and supported by convergent lines of argument. Thus, though theism is *ad hoc*, there is no reason in principle why it could not constitute a good explanation.

It has been argued that explanatory power, in the wide sense and not in Lakatos's limited sense, is the chief factor in acceptance of theories, but that this is qualified by two factors acknowledged by realists, namely that the theory be supported by convergent lines of argument and that it be plausible. But, it may be objected, this is still too vague, for the principles of theory choice which I have enunciated will not guarantee agreed solutions to the problem of assessing the worth of competing theories. We have already seen, however, in our earlier discussion of Kuhn's work, that it is in the nature of scientific reasoning that there can be no such guarantee. There is no algorithm which will decide between scientific theories, though this does not compromise the rationality of science. Kuhn writes that 'In a debate over choice of theory, neither party has access to an argument which resembles a proof in logic or formal mathematics',[54] and McMullin takes up that theme in an important article 'directed to showing that the appraisal of theory is in important respects closer in structure to value judgement than it is to the rule-governed inference that the classic tradition in philosophy of science took for granted'.[55] So it is that we must see scientists as exercising the rationality of judgement, rather than the rationality of rules. An observation of Wisdom's concerning legal judgement, to which Mitchell gives wider application, is to the point here: 'The process of argument is not a *chain* of demonstrative reasoning. It is a presenting and representing of those features of a case which severally co-operate in favour of the conclusion.'[56]

The point has been that in inferring to the best explanation one is engaged in informal reasoning more akin to moral or legal reasoning than it is to the reasoning of deductive logic which was at

[54] 'Reflections on my Critics', 260.

[55] E. McMullin, 'Values in Science', Presidential Address to the Philosophy of Science Association, *Proceedings of the Philosophy of Science Association*, 2 (1982), 8–9. Newton-Smith stresses the importance of judgement at many stages of the scientific enterprise, and suggests that an account of the rationality of science must make room for skills like that of the wine taster, who may be unable to give a verbal account of the principles according to which an acceptable wine is blended. *Rationality of Science*, 232–5.

[56] Cited by Mitchell, *Justification of Religious Belief*, 45.

one time taken to be the paradigm of science. On this point, though it may not seem so, I am in agreement with Swinburne. Although his *Existence of God* is structured around Bayes's theorem, there is no doubt that this theorem relies on informal reasoning rather than supersedes it. In other words, notwithstanding that Bayes's theorem has the trappings of a formal method of decision between hypotheses, in the debates between theist and atheist it represents a means for encoding and so laying bare the decisions we have taken. The point may be obscured, however, by the apparatus of the theory.

Bayes's theorem states that the probability of a hypothesis *h* on evidence *e* and background knowledge *k* is a function of its prior probability $P(h/k)$ and its explanatory power $P(e/h.k)/P(e/k)$. That

$$P(h/e.k) = \frac{P(e/h.k)}{P(e/k)} \times P(h/k)$$

But the promise of Bayes's theorem to provide an exact mathematical measure of the support that the evidence lends to each theory depends upon a number of idealizations.[57] Suppose that we know the probability of the evidence given the hypothesis;[58] then if we are to arrive at a value for $P(h/e.k)$ we have to have some value for $P(h/k)$, the so-called 'prior probability' of the hypothesis. But the value of this term will itself depend upon our subjective (informally rational) beliefs about the world, for we lack a formal measure of prior probabilities.

So it is that the promised formal method necessarily becomes infected by a more informal sort of rationality. Some have denied this, holding that whatever 'priors' are assigned there will be a convergence as the evidence comes in. However, in the theistic case the evidence is not of this nature. It will not be 'coming in' in any sense and thus makes Putnam's general conclusion of undoubted applicability here:

If scientists with different prior probability functions will not come into

[57] Here I follow the account of the difficulties with Bayes's theorem given by Putnam in his *Reason, Truth and History* (Cambridge: Cambridge University Press, 1981), 189–93.

[58] This is easy enough where the hypothesis is cast in probabilistic form, or where the evidence is entailed by the hypothesis, but neither is likely to be the case in theistic arguments. To imagine that any piece of evidence is entailed by God's existence would be problematic given the perfect freedom which theists have usually wanted to attribute to God.

agreement until the phenomenon to be predicted has already taken place, or until millions of years have passed, then, in the short run, the fact that there is some mathematical guarantee of eventual convergence is useless; the trouble with long-run justifications is that the long run may be much too long.[59]

Thus in the sorts of cases that Swinburne considers, Bayes's theorem can only be used to express our judgements, not to help us reach them. Far from being a means whereby disputes may be settled concerning the validity of a particular argument, it is more likely a means whereby one encodes one's prior judgements in a more technical form. That is not to say that those judgements themselves lack rationality, only that their rationality is neither derived from, nor displayed by, the theorem. Whether our inference is to theism or to some other hypothesis, we have no choice but to rely on informal reasoning. And this is so in science as much as in religion.

4. The Place of Simplicity

This chapter has been concerned with consideration of objections 'in-principle' to the idea that theism may be justified by an inference to the best explanation. The first objection doubted that theism could ever be an explanation, the second that theism could ever exemplify the qualities of a *good* explanation. These two charges required attention to the nature of explanation as well as discussion of criteria used in judging whether a hypothesis is worthy of belief. In particular, consideration and rejection of a stipulation against *ad hoc* theories led to an elucidation of the criteria of convergence and plausibility which function as checks on the main determinant of acceptability, namely, explanatory power. It was seen that there was no reason, in principle, why theism could not qualify as a good explanation.

One factor from the list given by Harman of the criteria to be met by a good explanation has thus far gone unmentioned, and that is simplicity. It remains then to ask whether theism is a theory which might be said to be simple. A prior question might arise, however, and that would be as to the place of this criterion in theory choice. Do we have any reason to prefer simple theories and to hold that they are more likely true?

---

[59] *Reason, Truth and History*, 191.

That simplicity is one of the guiding principles of scientific reasoning is sometimes regarded as a commonplace. We must, however, be clear about what is being claimed. What is in dispute is not whether a theory's simplicity is a reason for preferring it to a rival, but whether its simplicity constitutes a reason for believing it. Unless we are careful to make that distinction, we may mistake 'pragmatic preferring' for 'epistemological preferring'.

No one would want to deny that simplicity provides pragmatic grounds for preferring one theory (or at least—one formulation of a theory) to another. For example, take the inverse square law:

$$G = g \frac{m_1 m_2}{r^2}$$

Though there may be no difference in fit with the data between this law and one which raised r to a power such as 2.000000000001, our preference for the former style may be based on the reason that we find it easier to employ in calculations. But this preference should not be confused with an epistemological preference.[60] Our favouring a theory as a more amenable tool for calculation, to be preferred as a basis for research whilst it remains empirically indistinguishable from its rivals, is quite different from taking the simplicity of a theory as evidence of its truth or approximate truth.

Preferring a theory for its simplicity may not always be what it seems, here because the preference may be pragmatic rather than epistemological. But it may also be that where a theory is indeed *believed* on grounds of simplicity things may not be as they appear. For what is sometimes meant by 'simplicity' is not really simplicity at all. Thus although the preference is an epistemological one, it is not a preference founded on simplicity. So, to take one example, though J. L. Mackie suggests that simplicity is a guide to choice between theories, his suggestion must be distinguished sharply from that implicit in certain methodologies, such as Swinburne's in *The Existence of God*, for Mackie attaches a limited meaning to 'simplicity'.

The context of Mackie's suggestion is the consideration in his *Problems from Locke* of Locke's commitment to realism.[61] Mackie is particularly critical of an exegetical argument which runs as

[60] Newton-Smith, *Rationality of Science*, 231.
[61] J. L. Mackie, *Problems from Locke* (Oxford: Clarendon Press, 1976), chs. 1 and 2.

follows: though Locke appears to be a realist (for his distinction between primary and secondary qualities presupposes realism), he cannot have been committed to a view which can be refuted by a first year undergraduate, the refutation being that if we perceive nothing but ideas then we cannot know that some of these ideas are more like reality than are others.[62] But according to Mackie, such exegesis amounts to begging the question, for it presupposes rather than shows that there is no good argument for realism about material objects. In defence of interpreting Locke as a realist, Mackie proposes an argument for realism relying on simplicity, and this argument I shall briefly consider.

One solution to the 'veil of perception' problem which motivates the questionable exegesis is, of course, a naïve realism in the manner of Johnson. It regards 'perception as a mode of awareness that is so direct as to be self-guaranteeing and unproblematic, as simply *giving* us real external objects.'[63] But if this is unsatisfactory, are we left to embrace a Humean phenomenalism which allows merely for appearances as the limit of our ontology? No, because it is a mistake to suppose that there is no way out of the circle of ideas. We can argue that 'the real existence of material things outside us is a well-confirmed outline hypothesis, that it explains the experiences that we have better than any alternative hypothesis would, in particular better than the minimal hypothesis that there are just these experiences and nothing else.'[64] If we do not suppose that there are material objects, Mackie claims that the orderliness of our experience would itself become mysterious. Whilst the realist hypothesis explains why what we call two successive observations of the same object have the character of being continuous one with another, for the phenomenalist the repetition of remarkably similar groups of experiences would be entirely fortuitous. As Mackie puts it:

What is essential in this outline hypothesis is that it fills in gaps in things as they appear, so producing continuously existing things and gradual changes where the appearances are discontinuous. Its resulting merit is a *special sort* of simplicity, the resolving of what would on a rival, phenomenalist, view, be quite unexplained coincidences.[65]

---

[62] Mackie suspects that the argument has wide currency, but attributes it in particular to A. D. Woozley; *Problems from Locke*, 8.

[63] Ibid. 70.

[64] Ibid. 64.   [65] Ibid. My italics.

It is in the context of expounding an argument for realism concerning material objects then, that Mackie makes his appeal to simplicity, and in justifying this move refers to reasoning of a similar nature in certain astronomical debates. But it is important to see that in the argument I have set out above, and in the example from the scientific realm, Mackie does not mean by 'simplicity' what is sometimes meant. Indeed Mackie talks of 'a special sort' of simplicity (to be distinguished from the various other sorts which have been thought to be meritorious) 'dramatically illustrated by the contrast between Copernican and Ptolemaic theories'. He writes:

The key difference between them is not, say, over the mere number of cycles and epicycles used. Rather it is this. In a Ptolemaic account of the motions of the planet Jupiter, for example, there will occur somewhere a cycle or epicycle with a period of 365 days. Similarly in the account of the motions of the planet Mars there will occur, quite separately and independently, a 365 days' cycle or epicycle. And similarly with each of the planets, as well as for the sun. But since these are all independent, the recurrence of the same period, 365 days, in different places is a sheer unexplained coincidence. On the Copernican hypothesis, however, these separate epicycles disappear into the single revolution of the earth about the sun: there is no longer any coincidence to be explained. It is this, I believe, that constitutes the real initial superiority of the Copernican hypothesis over the Ptolemaic, as distinct from its subsequent confirmation by the way in which it led on to the more complete astronomical theories of Kepler and Newton. It is simplicity of this sort that I would call *the elimination of unexplained coincidence*; and this sort of simplicity is of the greatest importance as a guide to the choice between alternative scientific hypotheses. And while the existence of material things is not itself what we would ordinarily call a scientific hypothesis, being rather a framework within which the particular hypotheses that we so describe are formulated, it can, when the question of its justification is raised, be seen to be like a scientific hypothesis and to have in its favour this same sort of simplicity, this same elimination of unexplained coincidence.[66]

---

[66] Ibid. 66–7. Lakatos and Zahar concur in this judgement on the Copernicus/ Ptolemy dispute: 'Why Did Copernicus's Research Programme Supersede Ptolemy's?', 185–9. They write (186): 'If an astronomer takes the Earth as the origin of his fixed frame, he will ascribe to each planet a complex motion one of whose components is the motion of the Sun. This is an immediate consequence of the Copernican model: a change of origin leads to the addition of the Sun's apparent motion to the motion of every other mobile. For Ptolemy this is a cosmic accident which one has to accept *after* careful study of the facts. Thus Copernicus *explains*

The main point I want to draw from this argument is that what Mackie means by simplicity is in fact an aspect of explanatory power and not a matter of simplicity *per se*. That is to say, far from simplicity being an independent means of deciding between two hypotheses over and above the criteria we have previously considered, it is in fact used here to highlight the explanatory power of certain hypotheses over rivals. Whereas phenomenalism can say that there are appearances and they are regular, realism can explain the regularity; whereas Ptolemy can say that the planets move according to laws L_1, L_2 and L_3, Copernicus can explain why the laws have a certain character. Copernicus and the realist explain aspects of the data which are just taken for granted by the alternative theory. They provide a reason for what is otherwise merely observed and said to be compatible with the theory.

This may not seem obvious, for we can easily overlook the fact that the preferred theories áre explaining more; not more data if we take that in the sense of observations of planets previously unknown or ignored, but rather relationships and patterns within the data which constitute the evidence. (Darwin's argument concerning the family resemblances demonstrated by morphology is an example of appeal to a pattern.) A pattern does not look like an extra datum, for we tend to imagine new observations being won by patient application of ingenious experimental technique, whereas a pattern may be discovered in an armchair. But that is by the way, for the claim of the superior theory is that the pattern is a part of the evidence which is explained by the new theory and not by the old. The better theory sees something new, which the other does not notice, or, if it does notice it, denies to be a pattern and asserts instead to be a mere coincidence. It is in this sense that the new theory possesses greater explanatory power.

Simplicity has been thought to be prevalent in scientific reasoning. In assessing the relevance of that point for the claim that simplicity provides a reason for judging a theory worthy of belief against a rival, we must be on our guard against a failure to note two facts about the role of simplicity in science: first of all, that simplicity is often used to ground a pragmatic, rather than an epistemological preference; and second, that what sometimes goes

what for Ptolemy is a fortuitous result, in the same way that Einstein explains the equality of inertial and gravitational masses, which was an accident in Newtonian theory.'

under the label 'simplicity' is often better thought of in other terms. In the example from Mackie I discussed, it is better thought of as an aspect of explanatory power. Failure to note these points fosters the impression that the principle under discussion,—namely that, other things being equal, a simple theory is more likely true than its rival—is built into the foundations and practice of modern science. The 'truism' may be supported by such illicit foundations, but the real question is whether there are good arguments in favour of the principle that simple theories are, other things being equal, more likely true.

In his book *The Existence of God* Swinburne adopts the maxim *simplex sigillum veri*, describing it as 'a dominant theme of this book'.[67] He is concerned 'to show . . . the crucial influence of the criterion of simplicity within science' and comments: 'If we are to adopt in our investigations into religion the criteria of rational inquiry which are used in science and ordinary life, we must use this criterion there.'[68]

Where is this crucial influence shown? Swinburne thinks that it is displayed in our deciding between theories equal in explanatory power. He writes that:

for any finite collection of phenomena there will always be an infinite number of different theories such that from each (together with statements of initial conditions) can be deduced statements reporting the phenomena observed with perfect accuracy (and it may be that but for some one of these theories these phenomena are not to be expected). The theories agree in leading us to expect what has been observed so far, but disagree in their subsequent predictions. We may wait for new observations of phenomena to enable us to choose between theories; but however many theories we eliminate by finding them incompatible with observations, we will always be left with an infinite number of theories between which to choose, on grounds other than explanatory power. If there are no theories of neighbouring fields with which some theories may fit better than others, the crucial criterion is that of simplicity. (And when our theories are very large-scale, there will be little in the way of theories of neighbouring field.)[69]

Put in Bayesian terms the probability of a theory is a function of its explanatory power, $P(e/h.k)/P(e/k)$, and its prior probability, $P(h/k)$. Where theories are equal in explanatory power, it is their prior

---

[67] *Existence of God*, 56.
[68] Ibid.
[69] Ibid. 55.

probability which is crucial to their assessment. In certain cases, where the possibilities are clearly defined, the prior probability of a hypothesis may be given a numerical value. Thus the prior probability of drawing a red card from a normal pack is a half. In other cases the prior probabilities can be given only comparative assessments, and here, alleges Swinburne, three elements enter in. The first is fit with background knowledge (what I have termed plausibility), the second is scope, and the third is simplicity. And it is simplicity which he regards as crucial in comparisons of large-scale theories, for the reason that scope is not very important in any case, and because there is little in the way of theories of neighbouring fields with which the hypothesis in question may fit.

Swinburne takes theism to be such a large-scale theory, and in assessing its prior probability excludes no factual information from the evidence, *e*. Hence all that is left in background knowledge, *k*, is what Swinburne describes as tautological background evidence. Thus $P(h/k)$ becomes a measure of the 'intrinsic' probability of *h*, and simplicity becomes a crucial factor in the discussion of theism.

Mackie finds the notion of assessing prior probability on the basis of nothing but logical truth, which is what Swinburne's method requires, incoherent.[70] Thus Mackie writes:

if the evidence, *e*, is to be that there is a complex physical universe, then the background knowledge or belief *k* must exclude this, and so will be able to include only logical and mathematical truths. What likelihood could the god-hypothesis have had in relation to these?[71]

Mackie's rhetorical question points to a problem: how am I to judge a hypothesis simple without reference to background knowledge? In relation to the question of the simplicity of theism, finding the notion of an agent possessing these powers in this degree simple surely depends on what I think about persons. Simplicity or intelligibility must not be confused with familiarity,[72] but even so cannot be entirely independent of what one knows about the world. And Swinburne seems to acknowledge as much when he writes that

---

[70] For a discussion of Swinburne's argument for reliance on simplicity and the objection which Mackie makes to it see ch. 2 of R. W. Prevost's 'Theism as an Explanatory Hypothesis: Richard Swinburne on the Existence of God', D.Phil. thesis (Oxford, 1985).

[71] *Miracle of Theism*, 99.

[72] Swinburne insists on this point in a reply to Mackie, 'Mackie, Induction, and God', *Religious Studies*, 19 (1983), 387.

'our understanding of when one theory is simpler than another is very much the product of our scientific and mathematical up-bringing'.[73] Yet in the case we have described, with the demand that one settle a theory's simplicity relative only to logical truths, Swinburne would seem to want it otherwise.

Maybe Mackie is mistaken and the notion of judging a theory simple against nothing but logical truth is coherent. But we are still left with our initial question which is this: what reason do we have for supposing that simplicity points to truth? Swinburne claims to show 'the crucial influence of the criterion of simplicity within science'. One supposes, then, that he intends some sort of inductive justification. Science has relied on simplicity in these sorts of cases—the ones he has isolated where $P(h/k)$ is the decisive term and where fit is irrelevant—and in this reliance has not been disappointed. Thus if we are looking for standards to be employed in theory choice these are the ones to pick.

But this could hardly be the argument Swinburne intends, for he gives few examples of where simplicity has been decisive in a theory's favour, and where choice governed by simplicity has been seen to pay off. In fact, in spite of the claim to show the crucial influence of simplicity within science, Swinburne's attention to historical evidence is negligible. The only case which receives more than passing reference is Kepler's troubles with the motions of Mars. And it emerges in a recent paper that Swinburne's argument in favour of simplicity in fact relies on another more fundamental principle, the principle of credulity.[74]

The principle of credulity holds that its seeming to us that something is the case is, other things being equal, good reason for us to believe that it is indeed the case. Now without this principle, Swinburne claims, we would make no epistemological progress. Further, he argues, we have a tendency to believe simple theories. Thus although he describes the principles of both credulity and simplicity as fundamental, it can be seen that it is the principle of credulity which enables us to trust our intuitions in regard to simple theories.

To question Swinburne's justification would raise many detailed issues which are beyond the scope of this chapter. But assuming

[73] *Existence of God*, 56.
[74] R. Swinburne, 'Simplicity and Choice of Theory in Science and Religion', unpublished paper.

that simplicity is an important criterion in theory choice, we should ask whether theism might be held to be a simple theory.

Far from the demand that a theory be simple constituting an objection to theism, it is Swinburne's contention that theism is particularly favoured by this criterion. According to Swinburne:

A theory is simple in so far as it postulates few mathematically simple laws holding between entities of an intelligible kind. By a theory postulating 'entities of an intelligible kind', I mean that it postulates entities of a kind whose nature and interactions seem natural to us.[75]

Here Swinburne proposes three measures of a theory's simplicity: the number, simplicity, and intelligibility of its laws. The number of the laws proposed by a theory is at least (intuitively) easy to understand. The third criterion is the naturalness of a law, which Swinburne glosses as a matter of how intelligible we find the entities which a theory presupposes. The second criterion of the simplicity of a theory relates to the mathematical simplicity of the laws it proposes. Of Newton's theory of motion, which Swinburne suggests meets his demands admirably, he writes that its four laws were 'of very great mathematical simplicity'.

Thus the law of gravitation stated that all material bodies attract each other in pairs with forces proportional to the product of the masses of each, m and m^1, and inversely proportional to the square of their distances apart (r)–mm^1/r^2. The relationship is simple because the distance is not raised to a complicated power (e.g. we do not have $r^{2.0003}$ or $r^{\log m}$), there is only one term (e.g. we do not have $mm^1/r^2 + mm^1/r^4 + mm^1/r^6$), and so on.[76]

It is Swinburne's contention that theism itself exemplifies an analogous simplicity. To take one aspect of that simplicity, he claims that theism 'postulates the simplest kind of person . . . there could be'.[77] God's simplicity is said to lie in his possession of capacities which are as great as is logically possible, for an omnipotent God is simpler than a limited one. A limit on his powers would cry out for explanation, whereas limitlessness does not.

My own intuitions, even when instructed by Swinburne's arguments, are far from feeling the force of this point, a force which may be further weakened, it seems to me, if one places the doctrine

[75] *Existence of God*, 52.
[76] Ibid. 53.
[77] Ibid. 94.

of God in a specifically Christian context by expounding it in Trinitarian terms. At the least one might say that to establish that a theory is simple, or simpler than a rival, is not always an easy task and one may be left with the sense that the 'conclusion that theism is the most simple alternative has a highly subjective air'.[78] However, I shall not press that point. Instead we might allow to Swinburne that judgements of relative simplicity are sometimes made with a fair degree of confidence, and put on one side the question, which properly belongs to the details of an argument for or against theism, whether the theistic conception of God makes it a theory which is simple over against its rivals. At least it has been suggested that if simplicity has an important place in the choice of scientific theories so too it may have a part to play in the justification of theism.

Conclusion

In this chapter I have tried to elucidate the criteria which govern the acceptance of scientific theories and to show that they are criteria which might, in principle, suggest the rational acceptability of theism. It seems, then, that there could be a significant analogy between the justification of science and of religious belief. In the next chapter I shall discuss the objection most often made to this suggestion.

[78] O'Hear, *Experience, Explanation and Faith*, 114. More generally, Laudan raises the question as to whether those who argue for simplicity are not guilty of what he terms 'semantic utopianism', that is, advocating a goal which is not specified sufficiently clearly for it to be recognizable. *Science and Values*, 52.

7.

Conclusions: The Problem of Evil and the Philosophy of Science

1. Introduction

A THEME of this book has been that taking science seriously is essential to the philosophy of religion. This is obviously the case if the question which was put in Chapter 1 is to be answered—does religious belief exhibit rationality comparable to the rationality of science? And not only is taking science seriously essential to the philosophy of religion, but doing so is far from inimical to the defence of Christian theism. For, in the first place, it suggests that the explanatory justification of Christian theism, envisaged by Mitchell and attempted by Swinburne, represents a pattern of argument typical of science. And further it shows, as was argued in Chapters 4, 5 and 6, that misconceptions about science may inspire 'religious' doubts as to the appropriateness of such a justification, and 'scientific' doubts about the analogies between the two. In fact consideration of science shows that a number of philosophical theories concerning the nature and justification of explanation, theories hostile to theism, are unwarranted.

By way of a conclusion which continues this theme, I want to discuss the character of a theistic answer to the problem of evil. Although I have eschewed the task of examining explanatory justifications of theism as such, I shall treat the problem in some detail. I am encouraged to do this by two of its features:

(1) the problem of evil is generally felt to present the most pressing and persuasive objection to belief in the existence of God. Indeed, a sense that the problem of evil is intractable will, for some, seem to justify a refusal to engage with the substantial case for theism. The rationality of theism cannot in fact be analogous to scientific rationality, they will contend, just because, whatever else is to be said for belief in God, there remains this outstanding problem which is never satisfactorily resolved. If an explanatory

justification of theism is to win a hearing, it is essential, it might be said, that this objection be dealt with.

(2) the problem illustrates, once more, the claim that reflection on science is, far from being destructive of religious belief, encouraging to particular, traditional modes of its defence.

This is not to say that I shall demonstrate the success of the theist's answer to the problem. Evil is more than a slight difficulty which some ingenuity can overcome and a good part of this chapter is devoted to stressing the forcefulness of the objection from evil. I shall, however, reject certain dissatisfactions with a possible answer. In this context it will be found that Kuhn's distinction between 'puzzle' and 'anomaly' is helpful to our understanding, and can bring us to a proper appreciation of the legitimacy of the concept of mystery in apologetics.

Believer, agnostic, and atheist will often centre their doubts on the problem of evil. Is it possible or is it plausible, they will ask, that a good, omniscient and omnipotent God allows there to be evil in the world? And if it is not possible, or possible but implausible, must we not modify the traditional concept of God, with whatever consequences that would have, or admit that God does not exist?

The twofold structure of the problem is important. Is it *possible* or *plausible* that God and evil coexist? The problem presents these two faces. Philosophers have been concerned, on the one hand, with a *logical* problem, that of reconciling the existence of a God who is said to be omnipotent and omniscient with the fact of evil in the world. If the reconciliation is impossible, then this constitutes a devastating attack on traditional theism. On the other hand, there is another problem which at one time or another will have troubled most believers, and this is captured in the *evidential* objection which has the following form. A good, omnipotent, and omniscient God would only allow evil in so far as it served some higher end. But there are many instances of suffering which appear to serve no such end. Therefore it is implausible to maintain that there exists a good, omnipotent, and omniscient God.[1]

Evil may be treated solely as a logical problem only by those who either believe themselves to be in possession of a formal (that is,

---

[1] This division of the problem into its logical and evidential aspects I borrow from S. J. Wykstra, 'The Humean Obstacle to Evidential Arguments from Suffering: On Avoiding the Evils of "Appearance" ', *International Journal for Philosophy of Religion*, 16 (1984), 73–93.

deductive) proof of the existence of God, or maintain, as does Plantinga, that belief in the existence of God is basic to their noetic structure and hence does not require defence by argument.[2] Now the first of these assumptions is not one shared by those who canvass an explanatory justification of theism. Their interest in such a justification will usually go hand in hand with a conviction that the deductive route is subject to insuperable objections. Thus for them a solution to the logical problem, though necessary, is not sufficient to meet the challenge of evil. They must have some answer to the evidential charge that there appears to be useless suffering which the God of traditional theism would not allow. Nor, even if they accept Plantinga's controversial views on the 'basicality' of belief in God, will they feel relieved of the need to seek an answer to the evidential objection, if only because of the responsibility they have towards those for whom the belief is not basic.

Plantinga's work in relation to the logical problem has charted the course that a satisfactory solution to that aspect of the problem of evil must take, and I shall start out by considering it. In this consideration of his work premisses will emerge which we shall need to carry over to a treatment of the common-sense problem. So by following this route we may hope to obtain a true sense of the character of the possible theistic answers to the problem of evil.

2. The Possibility of Evil

How is it that a good, omnipotent, and omniscient God allows there to be evil in the world? Either he is unable to prevent it or unwilling. If the former, then his omnipotence is called into question, if the latter, his goodness. Clearly this logical problem, pressed for example by Hume in his *Dialogues Concerning Natural Religion*, challenges the coherence of the theistic concept of God.[3]

Plantinga has sought a way out of the dilemma by considering the nature of omnipotence. Theists and atheists alike have often

---

[2] A. Plantinga, 'Reason and Belief in God', in *Faith and Rationality*, ed. A. Plantinga and N. Wolterstorff (Notre Dame, Ind.: University of Notre Dame Press, 1983), 16–93.

[3] For references to Hume and to more recent versions of the attack, see N. Pike, 'Hume on Evil', in *God and Evil*, ed. Pike (Englewood Cliffs, NJ: Prentice-Hall, 1964), 85–102.

held that an omnipotent God is constrained by no limits other than the most obvious and trivial logical ones (such as, that not even an omnipotent God can make a round square), which in any case are not really limits on God's power, but limits on what it makes sense to conceive of. Hence there is no possible world in which there are round squares, which is misleadingly reported in the claim that God is unable to make them.

Believing God's omnipotence to be 'limited' only by such qualifications, Leibniz concluded that God is able to create any possible world which he chooses, provided only that it is a world in which God exists. Now since God is omnipotent in this sense of being able to create any possible world, and is also omniscient and perfectly good, Leibniz was led to the conclusion that this is the best of all possible worlds.[4]

Were one to conclude that this is not the best of all possible worlds, it would follow according to Leibniz's reasoning that there is no omnipotent, omniscient, and omnibenevolent God. But the claim that this is the best of all possible worlds is not in fact defensible. There is a possible world in which I am freely more virtuous than I presently am, and that would be a better world than this one. Hence this is not the best of all possible worlds and God does not exist.

Plantinga refers to Leibniz's position as 'Leibniz's lapse'.[5] Against Leibniz, Plantinga argues that there are logically possible worlds which even an omnipotent God could not have created; in particular, possible worlds the character of which is determined by the actions of beings who are possessed of free will. Plantinga draws our attention to the point with the example of a certain Curley Smith, a fictitious mayor of Boston. Suppose that in a given situation Curley is offered a bribe, is free to accept it or reject it, and in fact accepts it. Now it follows that God is unable to create a world in which in exactly those circumstances (including that Curley is free) Curley rejects the bribe. Certainly he could make

---

[4] That this is a mistake, even given Leibniz's premisses, is often pointed out. The conclusion should be that this is a world than which there is none better; there may be several of equal merit. See Mackie, *Miracle of Theism*, 173. For doubts about the very concept of the best of all possible worlds, see Swinburne, *Existence of God*, 113–14. Having recorded these qualifications, I allow to Leibniz, for the sake of argument, the legitimacy of the concept.

[5] A. Plantinga, *God, Freedom, and Evil* (Grand Rapids, Mich.: Eerdmans, 1977), 44, and *The Nature of Necessity* (Oxford: Clarendon Press, 1974), 184.

Curley reject the bribe, but then that would be a different world from the one in which he *freely* rejected the bribe, for making Curley reject the bribe is not the same as making Curley, who himself freely rejects the bribe. And Plantinga holds that whilst one is within the power of an omnipotent God, the other is not. Thus the moral is that here is a possible world which it was not in God's power to create, and that there will be as many more as there are situations in which there are beings exercising free will. In other words, it lies in Curley's power, and not in God's, which possible world will become actual. God is *unable* to create certain possible worlds. Hence 'Leibniz's lapse', and the mistaken commitment to the thesis that this is the best of all possible worlds.[6]

The compatibilist critics of the free will defence, such as Mackie and Flew, deny that Leibniz's lapse *really is* a lapse for they deny that the freedom which is required for moral responsibility in any way limits God's action. Their argument rests on an analysis of the notion of free will which maintains that there is nothing incompatible between the possession of free will and determinism, and which shows that God could have made creatures who freely choose the good. But they have another argument too. Mackie would assert that even if Leibniz's lapse is a lapse and the compatibilist analysis is mistaken, 'free will' can only be understood as a matter of randomness. And if that is so, then even if there are worlds which an omnipotent God cannot make (that is, worlds in which Curley randomly declines the bribe), they are not worlds which include the best possible world—for there can be no justified moral preference for worlds in which a creature's behaviour is random. Thus the exposure of Leibniz's lapse is of no help to the theist.

Mackie leads us through the first objection in the following steps:

(1) Given that it is logically possible that on one occasion a person should freely choose the good, can it be logically impossible that all people should always do so? Mackie argues that we can say that it is logically impossible only if we hold that what is done freely, at some time must not be done. But it is obvious that to hold this would be fallacious, even if in the tradition of Aquinas's move in the third way: 'a thing that need not be, once was not'.[7] It has to

    [6] *Nature of Necessity*, 173–84.
    [7] St Thomas Aquinas, *Summa Theologiae*, ii, trans. T. McDermott (London: Eyre & Spottiswoode, 1964), Ia. 2, 3.

be allowed, therefore, that it is not logically impossible that all people should always freely choose the good.[8]

(2) Mackie then asks whether it might be logically impossible that persons be *such that* they always freely choose the good. Mackie thinks not:

On a determinist view, what agents choose to do results causally from what they antecedently are, and the ascription of freedom denies external constraints which would make their actions depend on something other than their natures; it may also deny certain internal, mental, conditions which would prevent their choices from being proper expressions of their natures. So what a determinist calls free choices flow determinedly from the nature of the agent, and it follows that if it is possible that men should always freely choose the good, it must be equally possible that they should be *such that* they do so.[9]

If, as Mackie contends, to be free is to be able to do as one chooses or to act according to one's nature, then there would be no problem for God. He could make creatures with natures such that they always take the moral course, and they would then be both free and good.[10] So to go back to Plantinga's example, God *could* create a world in which Curley freely takes the bribe or one in which he freely does not. He could have so designed him that he freely chooses the good. But if that is so, then it was open to God to create the best of all possible worlds and the claim that our wills are free could provide no explanation for the existence of evil.[11]

The incompatibilist finds this analysis of free will inadequate and thinks that freedom and moral responsibility are undermined by the claim that our actions arise necessarily from natures which are themselves causally determined. Mackie's freedom is no sort of freedom at all. What seems lacking in a deterministic world view, the libertarian will say, is self-determination and the notion that somehow my actions are essentially mine. The force of the incompatibilist's analysis of our notion of free will comes across in John Hick's discussion of the conditions necessary for a person to be said to love God.[12]

---

[8] *Miracle of Theism*, 165–6.

[9] Ibid. 166. See also J. L. Mackie, *Ethics* (Harmondsworth: Penguin, 1977), ch. 9.

[10] Augustine seems to have believed that the will was both determined and yet free. For Augustine's difficulties with the issue see, G. R. Evans, *Augustine on Evil* (Cambridge: Cambridge University Press, 1982), ch. 5.

[11] *Miracle of Theism*, 172.

[12] J. Hick, *Evil and the God of Love*, 2nd edn. (London: Macmillan, 1977).

In discussing Mackie's objections to the free will defence, Hick concedes to both Mackie and Flew that it is not logically impossible that we should be made such that we always freely choose the good.[13] But Hick suggests that the question upon which the discussion of evil has largely focused—namely, could God make us such that we freely choose the good—is in fact irrelevant. The key question for a Christian theodicy is whether it is 'logically possible for God so to make men that they will freely respond to Himself in love and trust and faith?'[14]

The distinction between these two questions is not clearly made by Hick and it is inconsistent to allow different answers for each. Surely what Hick means to concede is that it is not logically impossible that we should always freely choose the good or alternatively that we be made such that we always choose the good. He cannot consistently concede, however, that God could make us such that we freely choose the good.

The answer to what Hick takes to be the main question must, he thinks, be 'no'. It is not logically possible for God so to make creatures that they freely respond to him in love.

For it is of the essential nature of 'fiduciary' personal attitudes such as trust, respect, and affection to arise in a free being as an uncompelled response to the personal qualities of others. If trust, love, admiration, respect, affection, are produced by some kind of psychological manipulation which by-passes the conscious responsible centre of the personality, then they are not real trust and love, etc., but something else of an entirely different nature and quality which does not have at all the same value in the contexts of personal life and personal relationship. The authentic fiduciary attitudes are thus such that it is impossible—logically impossible—for them to be produced by miraculous manipulation: 'it is logically impossible for God to obtain your love-unforced-by-anything-outside-you and yet himself force it.'[15]

Hick is asking us to take a God's eye view of the issue, and to see what sense the compatibilist claim would have for God. Would God regard my actions as free merely because I do as I choose? That is to say, Hick asks whether God could see the responses of his creatures as embodying love towards him, if he made them so that they perform those outward acts which we usually take as

---

[13] *Evil and the God of Love*, 271.
[14] Ibid. 272.
[15] Ibid. 273. The quotation with which Hick ends comes from John Wisdom.

manifesting devotion. For God to do so, he suggests, would be tantamount to a hypnotist prizing the affection which one of his patients has for him when that affection was induced by hypnotic techniques:

> The contradiction involved here would be a contradiction between the idea of A loving and devoting him/herself to B, and of B valuing this love as a genuine and free response to himself whilst knowing that he has so constructed or manipulated A's mind as to produce it.[16]

The computer programmer will not be flattered by a computer programmed to pay its user lavish compliments. No more will God cherish the devotion of human beings he has himself made 'devoted' or 'loving'.

Hick's point is sound, and impossible to square with his prior concession to Mackie and Flew that it is not logically impossible that creatures should be made by God such that they always freely choose the good. It would seem that essentially the same difficulties arise for this claim as for the claim that God might so make us that we freely love him. From God's perspective, my charitable actions towards my neighbour, designed by him as their manufacturer, can no more be valued than that love towards him which flows from the same source. Of course, it is possible that God might so make us that we *feel* as if we behave freely, but that is not what Mackie and Flew ask for. What is demanded is not that I feel as if I freely choose the good, but that I actually do so. And here the problem of being made so to behave arises once more. For let me suppose that I am the subject of an act of great compassion. My perception of it will alter in so far as I imagine that it is the result of a situation akin to that envisaged where God makes us loving. God might make people such that they behave well towards me. He might even make them such that I perceive them (and they feel) as if they are freely good. But from the God's eye view there can be no making them such that they freely choose the good. Thus Hick's concession is anomalous.

With the incompatibilist, Hick rejects an analysis of 'freedom'— be it freedom to love or freedom to do good—which sees freedom as constituted by an ability to act otherwise had one so chosen.[17]

---

[16] Ibid. 275.
[17] For other arguments against compatibilism, see, for example, R. M. Chisholm, 'Human Freedom and the Self', in *Free Will*, ed. G. Watson (Oxford: Oxford University Press, 1982), 24–35 and P. van Inwagen, *An Essay on Free Will* (Oxford: Clarendon Press, 1983), ch. 4.

Rather, to be free in respect of an action is to be able to choose to act differently from how one does, in spite of the physical conditions remaining the same.[18] It follows that free will and determinism are incompatible.

Of course, to advocate incompatibilism is not to show that libertarianism—the thesis that humans possess free will—is itself reasonable, even if one denies determinism. Indeterminacy may enter into the production of human actions, but such indeterminacy is merely a necessary condition for free actions, not a sufficient one. It is at this point that Mackie would renew his opposition, not limiting his attack to setting forth an opposed analysis of freedom. For notwithstanding that Mackie has no particular faith in determinism, he still maintains that the case for free will founders.[19] He asks the question: 'What could count as a freedom that both is of supreme value (either in itself or in what it makes possible) and is incompatible with an agent's being such that he chooses freely in one way rather than another?'[20] The dilemma which Mackie seems to threaten is this: action which is incompatible with an agent's being such that he or she chooses this way or that can derive only from what might be termed 'contra-causal' freedom; that is from a will which is not subject to causal factors. But if something is not caused it is thereby random, and no value can be attached to random actions. Indeed, they are not properly spoken of as actions at all. So just as the incompatibilist denies that God can value determined actions, so also must it be accepted that God would have no regard for action which is random. And yet the libertarian has, so Mackie alleges, no notion of free will which does not collapse into randomness. For if freedom is not defined as the compatibilist defines it (as an absence of certain constraints) what exactly is it?

Mackie is right, as I have said, in arguing that the mere denial of determinism does not ensure *self*-determination. But he is mistaken to argue that the libertarian is committed to what he calls 'contra-

---

[18] Plantinga's definition is: 'If a person S is free with respect to a given action, then he is free to perform that action and free to refrain; no causal laws and antecedent conditions determine either that he will perform the action, or that he will not. It is within his power, at the time in question, to perform the action, and within his power to refrain.' *The Nature of Necessity*, 165–6.
[19] *Miracle of Theism*, 168. For an fuller argument against determinism see van Inwagen, *Essay on Free Will*, ch. 6.
[20] *Miracle of Theism*, 169.

causal' freedom. As van Inwagen notes, the phrase 'contra-causal' is ambiguous, yet suggests the sort of freedom someone might have if his or her acts were uncaused. However, the libertarian claims only that our actions are undetermined, which does *not* entail that they are uncaused.[21]

In trying to distinguish agent causation from event causation, Chisholm claims that if we are responsible agents then 'each of us, when we act, is a prime mover unmoved. In doing what we do, we cause certain events to happen, and nothing—or no one—causes us to cause those events to happen.'[22] Like Aristotle's prime mover unmoved, we cause our acts though nothing determines us to cause them. Thus against van Inwagen and Chisholm it cannot just be supposed that an undetermined action is thereby uncaused and Mackie is wrong to allege that the free will defender has to believe in what is a contradiction, namely an uncaused free act. All that is required is a belief in an undetermined free act.

Chisholm's account is, I think, theologically admirable. It pictures each human being as in the image of God, for the individual will mirrors God's in its exercise of autonomous, creative power. But what it gains theologically it might be thought to lose philosophically, for it is by no means a transparent account, which is what Mackie seems to demand. As Watson puts it: ' "Agent-causation" simply labels, not illuminates, what the libertarian needs.'[23]

This feature is not, however, unique to the particular account of free will proposed by Chisholm, all such accounts tending to lapse into mystery at the point where we are required to picture the operation of the will of a self-determined, free agent. Chisholm talks, as we have seen, of the unmoved mover. Hick writes of 'limited creativity', and van Inwagen acknowledges the mysterious nature of free will.[24] But that free will is mysterious is, in itself, no reason for Mackie to reject it, for as Chisholm stresses event causation is itself mysterious. He suggests that the analysis of causation is not in good shape generally and that only the complete indeterminist avoids the problem of analysis altogether. It would be

---

[21] *Essay on Free Will*, 14 and ch. 4.
[22] 'Human Freedom and the Self', 32.
[23] Editor's Introduction to *Free Will*, ed. Watson, 10.
[24] *Evil and the God of Love*, 276; *Essay on Free Will*, 149–52 and 216.

more satisfying if we could take our analysis further, but that we cannot ought not to be taken as a decisive objection, even if we cannot say what agent causation *consists in*, but only distinguish free will from other things with which it is tempting to confuse it, such as contra-causal freedom, compatibilist freedom, and so on. For this is in fact no special difficulty for agent, as opposed to event, causation.[25]

The theist has specified the sort of freedom theism presupposes, and has argued that if creatures are endowed with such freedom, then which possible world will become the actual world depends on those creatures and not on their Creator. In discussing Hick's objection to compatibilism, a reason for God willing there to be such creatures has emerged. For if God wishes to create in his own image and have a reciprocal relationship of love with his creation, then free will is an essential requirement.

We have been engaged in sorting out Leibniz's lapse, which involved the supposition that God is able to actualize any possible world. This supposition seemed to render insuperable the problem of logical consistency posed by Hume, and commits the theist to the view that this is the best of all possible worlds. A way out of the dilemma is offered by Plantinga's contention that there are possible worlds which God is unable to create: for example, possible worlds which depend for their character on the random actions of creatures, or on the actions of agents with free will.

As yet though, this argument does not account for the existence of moral evil. As Plantinga puts it, the fact that Leibniz's lapse is a lapse 'does not settle the issue in the Free Will Defender's favour'. For, he continues, 'we must demonstrate the possibility that among the worlds God could not have actualized are all worlds containing moral good but no moral evil.'[26]

Plantinga seeks to demonstrate this possibility in the following way. Let us suppose, he says, that an omniscient God has knowledge of future contingents; that is, knowledge of the future behaviour of his free creatures. He knows, then, that if placed in the particular situation envisaged, Curley Smith will accept the bribe.[27]

---

[25] Chisholm, 'Human Freedom and the Self', 31.
[26] *Nature of Necessity*, 185.
[27] Hick attacks Plantinga on the ground that his solution to the problem of evil depends on God's ignorance about the future: *Evil and the God of Love*, 368–9. But why Hick should think this it is difficult to say, since Plantinga plainly expresses the

Our advice to the Creator might be this: don't let the situation arise in which Curley is offered the bribe. But what, asks Plantinga, if Curley suffers from transworld depravity, where transworld depravity is the condition such that in every world God may actualize in which Curley is significantly free, Curley takes at least one wrong action? Again the answer is apparent: don't create Curley. But what, Plantinga asks again, if every creaturely essence suffers from transworld depravity? Then if free will is essential for moral goodness, there is no world which God can create which contains moral goodness and yet does not contain moral evil.[28]

With these speculations Plantinga claims to demonstrate the logical consistency of belief in the existence of a good, omniscient, and omnipotent God in a world which contains moral evil, and thus to rebut the philosopher's logical argument from evil against theism. (Plantinga has thoughts on the existence of natural evil too, but we need not consider these.) He denies that there can be a deduction from 'this is not the best of all possible worlds', or 'this is a world which contains moral evil', to 'there is no God'.

At this point, however, Plantinga's argument is unsatisfactory. In reply to the question why God did not decline to create Curley in those worlds in which he behaves badly, or why he did not decline altogether to create a being depraved in all possible worlds in which he is free, we are told that God faced a limited choice. Curley was such that he would sin in all those worlds in which he was free, and his fault is universal amongst creaturely essences, both natural and supernatural. And yet, as Mackie points out, no reason is given for supposing that God was faced with these restricted options:

how is it possible that every creaturely essence suffers from transworld depravity? This possibility would be realized only if God were faced with a limited range of creaturely essences, a limited number of possible people from which he had to make a selection, if he was to create free agents at all. What can be supposed to have presented him with that limited range?[29]

opposite presupposition; e.g. in *The Nature of Necessity*, 185–6, he writes that '. . . God knows in advance what Curley would do if created and placed in these states of affairs.' And in *God, Freedom and Evil*, 43, Plantinga states that God knows which of two conditionals are true, where one says that so and so will freely do *x*, the other saying that he freely will not.

[28] *Nature of Necessity*, 184–9.
[29] *Miracle of Theism*, 174.

Mackie's point is well made. The existence of moral evil is explained only if we allow that transworld depravity obtains. But to allow that it does would involve an implicit denial of omnipotence. That is to say, whereas it is a logical truth that an omnipotent God cannot determine the actions of free creatures, it is not a logical truth that an omnipotent God cannot make a creature who does not suffer from transworld depravity. To concede that God is limited in this latter way is to concede that he is contingently unable to fulfil his intentions, and so not omnipotent.

Plantinga's reconciliation of the existence of evil with the existence of the God of traditional theism does not work, but a qualification of his position seems more promising. Contrary to what Plantinga supposes, let us suppose that God has no knowledge of future contingents. This is not to deny his omniscience, for if future contingents lack a truth value there is nothing which God fails to know. An omniscient God can know only those propositions which are knowable, just as an omnipotent God can do only those things which are logically doable. Thus if future contingents lack a truth value, as Prior and others have argued, then they are not knowable and therefore it reflects not at all on God's omniscience that he does not know what will happen when that is dependent on the wills of his creatures.[30]

Now it follows from this position on future contingents that an omniscient and omnipotent God would, so to say, create in the dark. In other words it is not possible for such a God to know in actualizing a world with free creatures that he is actualizing a world in which there is moral good but no moral evil. The very act of creation brings with it the risk that his creatures may make choices which are evil. Thus there can be no deduction from evil to the non-existence of God, and the theistic claims are shown to be at least consistent.

---

[30] I apply here A. Prior's analysis of future contingents in 'The Formalities of Omniscience', *Philosophy*, 37 (1962), 114–29. J. R. Lucas argues in *The Freedom of the Will*, 71–7, that if God did not lack knowledge of future contingents, there could be no human freedom. 'If we take time seriously, we must believe that what is past—including the fact that at some date in the past God foreknew what I am going to do—is unalterable, and in that sense necessary; and then . . . it is necessary that I shall act in a particular way.' (74)

3. The Plausibility of Evil

The common-sense, evidential objection to the existence of God had, it will be recalled, the following form. A good, omnipotent, and omniscient God would only allow evil in so far as it served some higher end. But there are many instances of suffering which appear to serve no such end. Therefore it is not reasonable to assert the existence of the God of traditional theism. The fulfilment of Plantinga's limited purpose (namely finding propositions such as to render consistent the existence of evil with the existence of a good and omnipotent God)[31] is thus only half of the task for those who seek an explanatory justification of theism. They have not only to respond to those who think that the theistic claims are inconsistent, but also to those who, while conceding that they may be consistent, contend that they are implausible.

Plantinga's argument against Leibniz's lapse, suitably qualified, constitutes a necessary first step in a defence of theism against the problem of evil. But it is no more than a first step, for it establishes, by the claim that future contingents lack a truth value, only that the existence of evil is (as a matter of logical possibility) compatible with the existence of God as traditionally conceived. It does not suggest that the amount and character of evil in the world is at all plausible or likely given the existence of a God who has no knowledge of future contingents. Thus it is inadequate as a response to the evidential objection, as, G. N. Schlesinger, for example, has pointed out.[32] Of course, God works in the dark in creating beings with free will and one of his risks may turn out to be a bad one. But this leaves unanswered the question why, when the character of a particularly harmful creature becomes clear, God does not intervene so as to limit the harm that is being done. An appeal to the unknowability of future contingents really does not solve the problem.

A solution to the evidential objection remains outstanding, and the question is what form that solution should take. A natural assumption would be that it must consist in reflections on the good

---

[31] See Plantinga, *Nature of Necessity*, 186: 'the present claim is not, of course, that Curley or anyone else is *in fact* like this, but only that this story about Curley is *possibly* true.'

[32] G. N. Schlesinger, *Metaphysics* (Oxford: Blackwell, 1983), 56.

purposes which evil may serve, but Steven Wykstra in his 'The Humean Obstacle to Evidential Arguments from Suffering: On Avoiding the Evils of "Appearance" ', thinks that the rebuttal of the evidentialist charge is more easily accomplished.

According to Wykstra, the theist has no duty to try to spell out the justifying reasons which God may have for allowing evil in the world. All that has to be done is to insist that the sceptic's claim, that evil appears to have no justification, fails to satisfy what he terms the 'Condition of Reasonable Epistemic Access'. This Condition states that 'On the basis of cognized situation s, human H is entitled to claim "It appears that p" only if it is reasonable for H to believe that, given her cognitive faculties and the use she has made of them, if p were not the case, s would likely be different than it is in some way discernible by her.'[33] One of Wykstra's examples will make the point clear.

Suppose that I am searching for a table and look through a doorway into a room the size of an aircraft hangar, filled with all sorts of bulky objects such as bulldozers, dead elephants, and such like. Standing in the doorway and seeing no table, should I say, 'It appears that there is no table in the room'? This would be an extremely misleading statement and one which I ought only to make if I were entitled to suppose that, had there been a table in the room, I would have seen it. And since the aircraft hangar is stacked from floor to roof with junk of one sort or another, its appearing that there is no table in the room when I open the door is of no epistemic significance.

Carrying this uncontroversial point over to the evidential objection to the existence of God, Wykstra argues that we are entitled to maintain that it appears that there is no justifying reason for evil, only if we are confident that if there were such a reason we would be privy to it. But according to Wykstra such confidence would be unjustified. He writes:

We must note here, first, that the outweighing good at issue is of a special sort: one purposed by the Creator of all that is, whose vision and wisdom are therefore somewhat greater than ours. How much greater? A modest proposal might be that his wisdom is to ours, roughly as an adult human's is to a one-month-old infant's. . . . If such goods as this exist it might not be unlikely that we should discern some of them: even a one-month-old infant can perhaps discern, in its inarticulate way, some of the purposes of his

[33] 'The Humean Obstacle', 85.

mother in her dealings with him. But if outweighing goods of the sort at issue exist in connection with instances of suffering, that we should discern most of them seems about as likely as that a one-month-old should discern most of his parents' purposes for those pains they allow him to suffer— which is to say, it is not very likely at all.[34]

Wykstra's conclusion is that 'for any selected instance of intense suffering, there is good reason to think that if there is an outweighing good of the sort at issue connected to it, we would not have epistemic access to it.' Thus the sceptic's challenge loses its force and the claim that there appears to be no good reason for suffering is shown to be unjustified and of no epistemic significance.

This answer to the evidential problem maintains that our inability to conceive of reasons which might justify evil in the world is of no significance given the nature of God as we suppose it to be. But might not this answer be turned back on the believer, and so take the very ground from under the feet of one who proposes an explanatory justification of theistic belief? For that justification relies on our making sense of certain phenomena by positing divine purposes and intentions. An explanatory justification of theism would claim that certain events are best explained by the supposition that God brought them about. But can we attribute to God reasons for bringing about such things, at the same time as we declare ourselves unable to judge why God might have allowed suffering and evil? Can we claim significance for its appearing to us that God might will, say, that the world should be beautiful, at the same time as we refuse to admit the claim that there appear to be no justifying reasons for evil? Can the critic not use Wykstra's reasoning to insist that its appearing to us that God has a reason for doing certain things would give rise to a justified claim *only if* it were the case that if God lacked the suggested reason, we could reasonably expect to know that to be so? And does not Wykstra's claim about the nature of God show that we would not be justified in such an expectation?

Of course there is not always a symmetry between 'appearing' claims. Thus, though the claim that 'it appears that the sea has no bottom' is unwarranted when made half-way across the Atlantic, it does not follow that the claim that 'it appears that the sea is shallow' made two yards from the edge of Brighton beach is

---

[34] Ibid. 88.

unjustified. We have epistemic access in the one case and not in the other. But in the case of reasoning about God, there is a symmetry which leads us to a certain scepticism. If we cannot, in principle, draw any conclusions from our not perceiving reasons God may have for allowing evil, then no more should we draw any conclusions from our finding plausible reasons why, say, God might grant mystical experiences to certain believers. In this case its being said that our failure is of no significance casts doubt on our claimed success.

I have suggested that the advocate of an explanatory justification of theism could not in consistency accept Wykstra's criticism of the evidential objection. Moreover that is good reason not to accept it, for it threatens an assumption about the relationship between God and man which the explanatory justification with good reason presupposes. Wykstra's argument involves a stressing of the gap between God and man as grounds for our inability to comprehend God's purposes. But I would wish to argue that such a stress is untrue to our best understanding of the relationship between humankind and God. For Christians believe that God creates mankind in his own image and for himself, with God willing that we realize our potential by entering into a relationship with him. Of course it is believed too that the consummation of this relationship will involve a transformation of humanity, both intellectual and moral. But against this has to be set the claim that there is a continuity between the two states, and that the envisaged consummation is of a previously existing relationship with God, albeit a relationship of limited quality and insight. Hence if this is so, then what appears to us to be the case here and now about God and his purposes is not to be set aside as of no significance, for the intellectual and moral gap between God and his creatures is not absolute. The significance of what appears to us to be the case will be bounded, so that it could not be expected that if God has a purpose in allowing suffering we could always comprehend it in every case. But we ought at least to comprehend it in some cases, even if darkly. The implication of this is that Wykstra's quick way with the evidential objection cannot be accepted, and that there is an obligation on the theist to offer some general account of why God might allow evil and what purpose it might serve. It is the nature and limit of that account which will now concern us.

Here the theodicist must retrace steps already made, and draw on the conclusion of the logical argument, that God cannot *make* creatures good and loving by divine fiat, for to be good and loving requires that one come to these attitudes freely. The question may then be asked as to what conditions might foster God's purposes. What must our environment be like for it to encourage growth towards the ideal without compelling it? Answering this question may help us to explain why the world is as we find it. We shall start out from a consideration of the so-called 'absorption argument' and go on to further suggestions made by Hick.

The absorption argument claims to solve the problem of evil by drawing attention to conditions essential for the fostering of certain types of good. Assuming the existence of free will, the absorption argument maintains that the creation of particular traits of character depends upon an environment which contains evil. It takes to heart the important and uncontroversial point that evil is a prerequisite of certain types of good. There cannot be courage without danger, there cannot be compassion without want, nor can there be loyalty without trial. Indeed, much of what we value in human life is lacking in the 'utopia' of Huxley's *Brave New World*, a point nicely captured in a comment of Virginia Woolf's about characters in a novel: 'I like people to be unhappy because I like them to have souls.'[35]

In its strongest form—the form I wish to consider initially—the absorption argument claims that *all* the evil in the world is to be explained by its purpose of provoking free creatures to good. One way of interpreting this claim is to maintain that each instance of suffering is permitted only because it provokes a worthy response and thus aids the divine purpose of character building. But can the theist plausibly argue that there is just enough evil in the world to produce second-order goods such as compassion, courage, charity, etc., the very existence of which require the presence of these evils? To take this line consistently, as Schlesinger does, is a heroic course, given the most natural reading of the world's situation as containing a vast amount of hopeless, redundant suffering.[36]

The most celebrated challenge to the absorption argument is Mackie's.[37] He accepts that first-order evil may be the means

[35] *The Letters of Virginia Woolf*, ed. N. Nicolson, iii (London: Hogarth Press, 1977), 294. I am grateful to Dr Jane Wheare for this reference.
[36] *Metaphysics*, ch. 3.
[37] J. L. Mackie, 'Evil and Omnipotence', in *Philosophy of Religion*, ed. Mitchell, 92–104.

whereby second-order good comes into the world. But by the same token he contends that second-order good may be the root of second-order evil, that is to say, evil which aims to frustrate a noble response to suffering or is itself an ignoble response to first-order evil. In an early version of his argument Mackie stresses that the situation threatens the theist with an infinite regress. But this is not the main point in *The Miracle of Theism*, where he contends that there appear to be surplus first-order evils, not actually employed in any larger organic whole of moral goodness.[38]

Schlesinger has detailed objections to the regress argument, but since Mackie no longer makes much of the threatened regress I do not propose to discuss what is at issue between them on that point.[39] We should, however, pay attention to Mackie's other argument, that there are instances of manifestly unabsorbed evil, for there is more than enough first-order evil to do the job of provoking us to second-order good. Schlesinger puts the objection in the following terms: though the 'virtuous response to suffering solution' provides an answer to the question why suffering should exist in those cases where it brings forth a noble response, an explanation is lacking for the existence of suffering to which no one responds virtuously, and to which no one *could* respond virtuously-—when, for example, a child, morally unformed, falls prey to an evil in isolation from others who might be improved by the child's plight.

Schlesinger proposes to deal with such a case by maintaining that the possibility for a virtuous response persists through time. Though the child may have died alone it is open to us to learn of the event and lament it, or be roused to make sure that no such thing should happen again. Given that there is this open-ended possibility for response, Schlesinger concludes that 'It is not possible to have evidence that useless misery, that is, misery that never has and never will elicit a philanthropic response from anyone, has ever existed.'[40]

But this, of course, will not do. Schlesinger's thesis may be unfalsifiable, yet are we not practically certain that there are instances of suffering in the history of the world which have gone unrecorded and which will remain undiscovered? Schlesinger's way

---

[38] *Miracle of Theism*, 153–5.
[39] *Metaphysics*, 61–2.
[40] Ibid. 63.

of dealing with the assumption that there are cases of unknown suffering is reminiscent of Berkeley's against the conceptual propriety of 'matter', where Berkeley confuses the unconceived idea of a tree with the idea of a tree unconceived. Certainly I can have no knowledge of an instance of suffering which simultaneously I do not know about, but that does not prevent my having a reasonable belief in instances of suffering of which I have no knowledge. It is only if one finds this belief implausible that one will be inclined to accept the absorption argument as satisfactory.

We were prompted to consider the absorption argument by asking the question: 'What conditions are necessary for the fostering of God's purpose in the world?' The absorption argument stresses the instrumental importance of evil in inspiring us to a virtuous response. Whilst I have not wished to call in question that fundamental insight, I have argued that it is not sufficient to account for all the evil in the world unless we are prepared to accept that every grain of evil directly serves a higher end in provoking good responses. It is the implausibility of this claim which leads me to abandon the absorption argument in the stark form advocated by Schlesinger.

A datum from which any theodicy must proceed, if our arguments thus far are valid, is the existence in the world of hopeless evil and suffering, or, more exactly, the existence of evil which seems to serve no specifically identifiable part in provoking others to virtuous responses. The theodicy proposed by Hick takes account of this point and accommodates it within the purposes of God. It differs from the absorption argument in that whilst it adopts and builds on the insight that evil is conducive to soul-making, it allows that the level of evil in the world is more than can be accounted for by virtuous responses. Though Hick's is an absorption argument of sorts, it does not say that every evil which is allowed contributes to a specifiable good response. There is useless evil, but it maintains that an environment which allows that is the only environment in which God's purposes may be fostered. An account of Hick's 'Irenaean theodicy' serves to elucidate these points.

The fulfilment of the end of humankind is found in the individual's acceptance of the love and goodness of God and his or her transformation into God's likeness. For someone to come to this likeness is for them to love God and will what God wills. But to share in God's life requires a free response to God; that was the point of the argument of section 2 of this chapter, which established that a genuine response to a person must arise freely if it is to be valued. Now if such are the purposes which God has for his creation, what conditions must obtain in the world for it to be conducive to this end? How is the life for which we are made to be fostered?

The first point Hick makes is that the demand for freedom, which God's purpose involves, in turn necessitates that humankind be set at 'an epistemic distance' from the Creator, in separation from a God who must be hidden. For to be free to come to know God, his creatures must not be overwhelmed by a presence which would jeopardize their 'cognitive freedom', but must be placed in a natural environment. This occurs through evolutionary development, with its implication that human existence is continuous with the existence of this world, not obviously indicative of another. Consequently the human outlook is naturalistic, world-centred, or 'fallen' as the traditional language would have it, and the prevalence of moral evil is a predictable concomitant of this continuity with the natural world.[41]

Hick further claims that the divine purpose for creation makes evident the need for natural evil in the world. For the environment to serve its purpose there must be such evils to be overcome. With this requirement Hick's theodicy adopts the insights of the absorption argument. The point is stressed that the present world is more conducive to the elevation of the soul than would be a quiet paradise.[42]

Even granted these two points, Hick allows that the actual level and character of suffering in the world will be seen as problematic. On this score he makes three further points. First, he offers a Kantian observation that the randomness of suffering is essential if the moral life is not to be undermined. If the good always prospered and the wicked always suffered then there would be no doing of good for the sake of doing good.[43]

Second, he asks rhetorically whether some of the suffering in the world is not just too great to serve any useful purpose. He suggests that we imagine that the worst suffering is eliminated, and what is

---

[41] *Evil and the God of Love*, 281–91.
[42] Ibid. 322–7.
[43] Ibid. 335.

then worst is eliminated, and so on until we reach an acceptable level, say an itch as the limit of physical pain and a mild regret as the limit of mental pain. He then claims that this will be a world in which there is no opportunity for real moral and spiritual growth, for there is no possibility of hurting anyone, no possibility of doing anyone good, and no place for the higher virtues which arise through a solidarity in, and sharing of, the suffering of others. As Hick puts it:

there is, I suggest, a deep connection between morality and suffering such that a world without the possibility of real—and therefore unacceptable—suffering could not be inhabited by morally growing beings. We may be tempted to feel that if morality and evil go together in this way, God should not have created moral beings. But to reject this temptation is to begin to see the possibility of the Irenaean type of theodicy.[44]

And again he writes:

In order to be an environment in which they can grow as moral and spiritual persons the world need not of course contain the particular pattern of hazards and challenges which it does contain. But it would need to contain either this or some other set; and whatever set this might be would inevitably seem both arbitrary and excessive to those who have to live in that world.[45]

There is a point that we should add to this train of thought regarding the existence of evil unmitigated by any good, a point which attempts to answer the objection to the absorption argument taken in isolation. It is that the existence of unmitigated and mysterious evil is not inimical to God's overall purpose. His answer above is partial, Hick would admit, for the sheer scale of evil in the world leaves doubts in our minds. But he is led to say this:

I do not [have a] theory to offer that would explain in any rational or ethical way why men suffer as they do. The only appeal left is to mystery. This is not, however, merely an appeal to the negative fact that we cannot discern any rationale of human suffering. It may be that the very mysteriousness of this life is an important aspect of its character as a sphere of soul-making. Perhaps, as the Cambridge Platonist Ralph Cudworth thought, 'it is in itself fit, that there should be somewhere a doubtful and cloudy state of things, for the better exercise of virtue and faith.'[46]

[44] 'Remarks', in *Reason and Religion*, ed. Brown, 127.
[45] *Evil and the God of Love*, 374; see also 378.
[46] Ibid. 333–4.

Hick would say that the creation of cognitive distance requires that mankind be set in a natural environment which itself is ambiguous in the witness it bears to God, and so allows for the existence of evil. Further, the evils which are prevalent in the world are essential to the creation of second-order goods. That is to say, the part of God's purpose which consists in the building of moral character is served by these evils. Add to these the point that any level of suffering which is such as to serve these ends would seem problematic, and that the mystery which remains when these observations are finished can itself be understood as contributing to the overall suitability of the world for God's purposes. God could not make the creation as he wills it to be. He could only foster its growth towards intended fulfilment. It is this purpose which, argues Hick, accounts for the character of the world as we find it. We do not have a degree of understanding which explains the actual incidences of evil that we experience. They remain mysterious. But, Hick argues, the fact that we do experience evil is itself not left inexplicable. We do not have an explanation of this or that evil, but we do have an explanation of why the world is so—why it is, if you like, that we do not have the particular explanations.

Let us take stock of the responses to the problem posed by the existence of evil in the world. Briefly, God may be thought to act in creation according to a purpose which constrains the means which may be employed in achieving it. That purpose stems from God's valuing humanity as capable of entering into a relationship of love with him, responding to his initiative. And the constraint is that if this is God's purpose for his creatures, he cannot make a world in which they must respond to him. To suppose that he can do so is to be guilty of Leibniz's lapse in believing that God may create any possible world. In fact, there are certain logically possible worlds which an omnipotent God is unable to create, such as worlds whose character is in part determined by the exercise of free will.

As a preliminary step, this criticism of Leibniz is essential, but it does not in itself establish the free will defence. It shows that there are worlds which an omnipotent God cannot create, but not that this single limitation explains the existence of evil. That is to say, it does not yet make clear why God was limited to creating only worlds which contained evil. Plantinga makes the limitation on God more vexing by appeal to the notion of transworld depravity, but whilst the constraint posed by free will is a logical limit, this one is contingent, and has to be rejected as entirely inexplicable.

Another way in which the limitation may be made more substantial is by arguing that God lacks knowledge of future contingents. I have contended, however, that though this lack is essential to human freedom, it is insufficient to explain the existence of evil. It is true that God creates in the dark so that moral evil may occur against his will, but that would account neither for God's allowing the continued exercise of a will as solidly corrupt as was Hitler's, nor for the prevalence of natural evil. The absorption argument goes further and proposes that all the evil in the world can be understood as providing an opportunity for worthy responses. This argument proves unable, however, to deal with the existence of evil, which clearly offers no such possibility. None the less Hick's theodicy builds on the same foundations as the absorption argument, and provides a fuller understanding of the proper state of a world in which creatures are free and yet have the opportunity and encouragement to develop as God intends. As against the strict form of the absorption argument it admits that there is suffering which appears to serve no good end, but insists that this is suffering which, within the mystery of evil, can be seen as conducive to God's plan.

4. Anomalies, Puzzles, and the Propriety of Mystery

According to traditional Christian belief the purpose of God's creation is that men should finally enjoy a communion with one another and with God which fully satisfies their hearts and minds, and the present world, with its suffering and its opportunities for moral evil, provides the only sort of environment in which men could develop the virtues needed to sustain and enjoy that status. Believing this and believing also, as Christians, that God has involved himself in the suffering of the world and, in so doing, shown how it may be transmuted, believers claim that they have been given some insight, however incomplete, into the mystery of evil.[47]

Mitchell's outline of the traditional Christian response to the evidential problem of evil serves to suggest that Hick's theodicy would need to be taken further if it is to be fully convincing. It would need not only to develop an understanding of life after death (which is of course a presupposition of the 'training ground' view),

---

[47] Mitchell, *Justification of Religious Belief*, 10.

but also to give an account of the doctrine of the incarnation and of its significance.

My purpose, however, in offering a sketch of a possible answer to the evidential problem was not so much to display its full power, as to draw attention, as does Mitchell, to the character of the answer as involving an ineliminable element of mystery. And for that purpose a sketch is sufficient; for it notes Hick's admission that actual occurrences of evil, both moral and natural, are themselves mysterious, even though that there should be such evil and that it should be mysterious is itself understandable given the overall purpose which God has for his creation. And it notes also that within the theodicy there lie other mysteries. There is the mystery of the nature of free will and agent causality to which I have already referred, and there is the mystery of life after death and the mystery of incarnation mentioned by Mitchell. All in all, the theodicy's character is well expressed in Mitchell's cautious and qualified claim on its behalf that believers 'have been given some insight, however incomplete, into the mystery of evil'.

Could the essentially incomplete and mysterious nature of this answer be a ground for ultimate dissatisfaction with it? Might it be maintained that the theist has given an improper and inadequate reply to the difficulties which animate the enquirer, a reply which is vitiated by its likeness to a promissory note? Is 'some insight' good enough? And if we hold that it is, have we not in effect given up the search for a defence of theism which might be as rationally compelling as a defence of a scientific theory?

My response to such feelings of dissatisfaction is to insist that the element of mystery in this theodicy is not, in itself, ground for rejection of theism. Nor does it preclude the possibility of theism being such as rationally to compel assent, any more than the same sort of mystery would deprive a scientific theory of its compelling character. Rather I would argue that Kuhn's critique of falsificationism makes room for mystery in scientific theories, and by analogy legitimates, in principle, the residual mystery in the theist's explanation of evil. In making this point I shall employ and sharpen a distinction between puzzles and anomalies suggested by Kuhn.

Kuhn's distinction between puzzles and anomalies is a response to the naïve falsificationism which some found in Popper's work and which I discussed in Chapter 5. The central feature of this naïve falsificationism is its contention that every difficulty for a theory amounts to a refutation. In reply to that contention Kuhn points out: 'If any and every failure to fit were ground for theory rejection, all theories ought to be rejected at all times.'[48] Kuhn shows that the practice of scientists is rather to treat difficulties with a promising theory as puzzles to which answers will, in due course, and with sufficient application, be forthcoming. As with a crossword puzzle, failure to find a solution, rather than suggesting that there is something wrong with the crossword, reflects on my abilities as a solver of such puzzles.

Scientists have always understood the need for a relatively tolerant attitude towards difficulties, as the example discussed in the last chapter suggests. For Darwin commends his theory of evolution by natural selection in *The Origin of Species* well aware that it is far from being perfect and free from puzzles. How can the extreme perfection of organs and instincts be produced by successive minor modifications? Why is the geological record so sparse when allegedly there have been countless generations of flora and fauna? How can the attribution of a great age to the earth which the theory presupposes be justified, when it goes against (then) current physical theory? How does the mechanism of inheritance work? As Kuhn maintains and Darwin's theory illustrates, even the best theories come replete with difficulties which are treated as puzzles.[49]

Problems for a hypothesis arise, then, not with the mere instance of a puzzle or a difficulty which cannot be fully comprehended by the theory. The falsificationist's simple view will not do, for theories cannot be rejected merely because they seem incomplete or fail to fit perfectly. But on the other hand, if we are to avoid excessive appeal to mystery, which is properly feared as a block to critical scrutiny, something more must be said. We cannot merely conclude that mystery is permissible without justifying the suspicions of the naïve falsificationist that, as Lakatos puts it, all this talk of puzzles and anomalies 'is a dishonest euphemism for counterevidence'.[50] And even where there is no suspicion of dishonesty we should still be on our guard against, for example, the assurance that 'the Holy Trinity is not a dark mystery or an

---

[48] *Structure of Scientific Revolutions*, 146.
[49] We might also note that the rational realism defended in Chapter 3 above treats the analysis of the concept of verisimilitude as such a puzzle.
[50] 'Falsification and the Methodology of Scientific Research Programmes', 120.

enigmatic mystery, it is a bright and dazzling mystery, one which attracts us by its very splendour'.[51] For mystery can seem like a cloak behind which to hide incoherence and counter-evidence, and unless we can provide some answer to the question of when puzzles become anomalies we run the risk that the suspicions which properly attach to some appeals to mystery will attach themselves, uncritically, to them all.

According to Kuhn, a difficulty for an established theory is *usually* regarded as a puzzle, which is to say that it is tolerated without its endangering the status of the theory in question. But the very same difficulty may change its significance from that of a puzzle to that of an anomaly, where 'anomaly' means refutation or counter-instance. It may do so where the difficulty seems to drive at the very heart of the theory, or is long outstanding, or is itself one of a proliferation of difficulties. What used to be tolerated is now cited as a reason for abandoning the theory. But, as Kuhn notes, 'even the existence of crisis [that is, fundamental or long-standing questioning of the paradigm] does not by itself transform a puzzle into a counterinstance.'[52]

It may seem that Kuhn does not really help us. Is it not possible to be more precise? Kuhn maintains that there is no 'sharp dividing line' between puzzles and refutations or counter-instances.[53] And to the question 'When does a difficulty for a theory merit the scrutiny which an anomaly deserves but which a puzzle doesn't?', he thinks that 'there is probably no fully general answer.'[54]

Should we not ask for something more than Kuhn's vague guidance? Can we not demand rules by which to differentiate puzzles and anomalies? If, however, theory choice is essentially a matter of judgement and cannot be determined by rules as I have maintained in previous chapters, then the demand that the choice be more precisely governed is one which cannot be satisfied. Let me refer once more to the important article by Ernan McMullin entitled 'Values in Science' in which he argues that 'the appraisal of theory is in important respects closer in structure to value judgement than it is to the rule-governed inference that the classic tradition in the philosophy of science took for granted.'[55] If

[51] A. and R. Hanson, *Reasonable Belief* (Oxford: Oxford University Press, 1980), 184.
[52] *Structure of Scientific Revolutions*, 80, and ch. 8, *passim*.
[53] Ibid. 80.
[54] Ibid. 82.
[55] McMullin, 'Values in Science', 8–9.

McMullin is right, dissatisfaction with Kuhn's vagueness, far from constituting a criticism of his position, merely expresses a hankering after the certainties of positivism. For it may be that we can say no more than that the transition from puzzle to anomaly occurs when it is judged that the particular successes of a theory are outweighed by its lack of suggestiveness in dealing with those problems which remain outstanding, problems for which there are competing and more promising alternative solutions. Here there are no rules to be offered for the choice which is involved, a choice which will require the exercise of judgement in the weighing and balancing of alternatives.

The nearest we can get to a formula would be this: that the greater a theory's explanatory power and plausibility, the more will its failure to resolve the mysteries associated with it be tolerated. It is a formula to which Darwin was sensitive in dealing, for example, with doubts about how a highly complex and adaptive organ or instinct can be produced by successive modifications, which in their early stages may not seem in the least advantageous. He writes:

He who will go thus far, if he find on finishing this treatise that large bodies of facts, otherwise inexplicable, can be explained by the theory of descent, ought not to hesitate to go further, and to admit that a structure even as perfect as the eye of an eagle might be formed by natural selection, although in this case he does not know any of the transitional grades. His reason ought to conquer his imagination; though I have felt the difficulty far too keenly to be surprised at any degree of hesitation in extending the principle of natural selection to such startling lengths.[56]

Darwin makes a valid move in his argument with the theory of special creation, and in theism's argument with competing world views the same move cannot be precluded out of hand. The theist is entitled to claim that the mystery which surrounds the problem of evil is balanced by the explanatory power of the theory in other fields. The difficulties should be stated as clearly as possible and the mystery probed as deeply as it can be. Then it may be possible to suggest the direction which an answer might take and so claim that evil is not the sort of puzzle that deserves the status of an anomaly. Without prejudicing an explanatory justification of theism, one cannot rest content with the mere observation that mystery is not necessarily fatal to a theory. But having done what can be done in containing the mystery, the theist is entitled to bid the doubter

[56] *The Origin of Species*, 218–19.

weigh the explanatory gains against the theoretical difficulties and problems. Whether the difficulties are such that theism should be rejected will turn, then, on the larger question of the explanatory power of theism in relation to its range of evidence: the existence and order of the universe, the allegedly providential and miraculous character of the history of Israel and of the Church, the occurrence of religious experience and of saintliness amongst the faithful, and so on.

The apologist's judgement is that theism has so much to commend it that we should treat the difficulties as puzzles. Theism has a resonance with reality, a fit with experience, so it is claimed, which leads the theist to say that the problems in the theistic hypothesis are not ultimately detrimental to it. Its present explanatory power holds the promise of a future, unambiguous confirmation. Words Darwin used of his own work, the theist will borrow to commend the apologetic endeavour:

a crowd of difficulties will have occurred to the reader. Some of them are so grave that to this day I can never reflect on them without being staggered; but, to the best of my judgement, the greater number are only apparent, and those that are real are not, I think, fatal to my theory.[57]

[57] *The Origin of Species*, 205.

8.

Afterword

THE remarks from Hume with which I began, were intended by Hume to be ironic.[1] But many religious believers today would be ready to accept them without demur. Thus it is often from this quarter that one encounters the greatest resistance to the suggestion that an explanatory justification of a scientific nature may be appropriate to the defence of Christian theism. Where this resistance relies not upon one of the detailed objections which I have already considered, I sense in it the popular conviction (which bears no relationship to Phillips's account of Wittgenstein) that argument is irrelevant to religious belief. For this conviction may contribute to an objection to the proposed analogy in the following way:

1. If the analogies between scientific and religious reasoning were really significant, then people would be argued in and out of faith.

2. But they are not.

3. So there must be an important disanalogy.

If the objector is prepared to speculate as to what this important disanalogy might be, the suggestion is often encountered that to understand and accept a scientific argument is a matter of pure intellect, whereas that is not the case with the acceptance of theism. For theism's subject matter is different from that of physics or chemistry, let us say. Christian theism considers how the intentions of a person, God, may explain the existence and character of the world. So to understand and accept theism requires insight and sympathy concerning the motives of a person, and such insight is not achieved by argument.

It should be acknowledged at once that proposition 3, expanded in this way, points towards an important difference between most scientific and theistic reasoning. But unfortunately for the objection, this difference does not explain the point made by proposition 2, namely, that people cannot be argued into faith. For reasoning

---

[1] D. Hume, *An Enquiry Concerning Human Understanding*, ed. L. A. Selby-Bigge, 3rd edn. (Oxford: Oxford University Press, 1975), 131.

about the intentions of persons has the same form, if not the same subject matter, as scientific reasoning. Thus both the scientist and the theist look at a range of facts and ask an essentially similar question: 'what sort of laws and objects, or person, would account for this?' The reasoning of Darwin's *Origin of Species* does not differ in kind from the reasoning of Paley's *Evidences of Christianity* or his *Natural Theology*, the logic of which, writes Darwin, 'gave me as much delight as did Euclid'.[2]

Of course, how good I am at reasoning about motives and intentions may depend on what sort of a person I am, with what sort of experience. Imagine that as a juror at a trial I am asked to consider whether the action of which the defendant is accused is to be expected of a jilted lover. If I have never been in love myself or lack the imaginative power to enter into the emotions of someone who has been rejected, then I may feel at a loss. In a similar way, in an argument about the problem of suffering, a theist may suggest that evil can be a blessing. But if I have never experienced the evil in question I may find it hard to conceive of the hardships of poverty, for example, in such terms.

But even in this sort of case the argument does not come to a standstill, for no matter the difficulty there is usually a further recourse. There is always the evidence of what lovers or saints have said about rejection and poverty, which might enable someone to grasp what is involved. Reading certain novels or poetry would probably help. Argument here may be demanding of sympathy and understanding in a way which is not characteristic of argument in the sciences, but what we have here is still rational argument.

Proposition 3 does not explain what it is devised to explain, namely proposition 2. But in any case, 2 does not need explaining since it is false. People *are* argued in and out of faith, a phenomenon which is invisible only to those so imbued with popular versions of the naïve falsificationism discussed in Chapter 5 that they notice no arguments which are not knock-down arguments. Certainly there are no effective knock-down arguments available to the theist, for the subject is too vast and its ramifications too many for that to be possible. But then the same point applies to scientific theories such as Darwin's theory of evolution by natural selection, or Einstein's

---

[2] C. Darwin, *Autobiography*, in *Charles Darwin and T. H. Huxley: Autobiographies*, ed. G. de Beer (Oxford: Oxford University Press, 1974), 32.

theory of general relativity. Here argument works in a way more subtle than the objection conceives.

Wittgenstein writes: 'When we first begin to *believe* anything, what we believe is not a single proposition, it is a whole system of propositions. (Light dawns gradually over the whole.)'[3] For many believers their experience is certainly one of light gradually dawning over the whole. It may be that the proximate and outward cause of their coming to faith is something other than argument or reasoning: bereavement, acquaintance with conspicuous sanctity, or participation in the sacramental life of the Church. But whatever may be the immediate cause, for many an essential preliminary will have been acquaintance with a careful and patient apologetic which clears a path to faith by defeating the various objections which are put in its way and which, in setting out an explanatory justification of theism, helps the enquirer to perceive that pattern in experience which points to the existence of God. There is no reason in principle why such apologetics could not be as compelling as the apologies which are made for currently accepted scientific theories. And since it may provide the grounds for faith, its neglect is more than regrettable.

---

[3] *On Certainty*, para. 141.

Bibliography

Achinstein, P. *Concepts of Science* (Baltimore: Johns Hopkins University Press, 1968).

Aquinas, St Thomas. *Summa Theologiae*, ii, trans. T. McDermott (London: Eyre & Spottiswoode, 1964).

Austin, W. H. *The Relevance of Natural Science to Theology* (London: Macmillan, 1976).

—— 'Religious Commitment and the Logical Status of Doctrines', *Religious Studies*, 9 (1973), 39–48.

Barnes, B. and D. Bloor. 'Relativism, Rationalism and the Sociology of Knowledge', in *Rationality and Relativism*, ed. Hollis and Lukes, 21–47.

Bernstein, R. J. *Beyond Objectivism and Relativism* (Oxford: Blackwell, 1983).

Bieri, P., R. Horstmann and L. Krüger, eds. *Transcendental Arguments and Science* (Dordrecht: Reidel, 1979).

Bloor, D. *Wittgenstein: A Social Theory of Knowledge* (London: Macmillan, 1983).

Boyd, R. 'The Current Status of Scientific Realism', in *Scientific Realism*, ed. Leplin, 41–82.

—— 'Lex Orandi est Lex Credendi', in *Images of Science*, ed. Churchland and Hooker, 3–34.

—— 'Realism, Underdetermination, and a Causal Theory of Evidence', *Noûs*, 7 (1973), 1–12.

Braithwaite, R. B. 'An Empiricist's View of the Nature of Religious Belief', in *The Philosophy of Religion*, ed. Mitchell, 72–91.

Brown, S. C., ed. *Reason and Religion* (London: Cornell University Press, 1977).

Brümmer, V. *What Are We Doing When We Pray?* (London: SCM, 1984).

Carl, W. 'Comment on Rorty', in *Transcendental Arguments and Science*, ed. Bieri, *et al.*, 105–12.

Chisholm, R. M. 'Human Freedom and the Self', in *Free Will*, ed. Watson, 24–35.

Churchland, P. M. 'The Ontological Status of Observables: In Praise of Superempirical Virtues', in *Images of Science*, ed. Churchland and Hooker, 35–47.

—— and C. A. Hooker, eds. *Images of Science* (Chicago: University of Chicago Press, 1985).

Cook, J. W. 'Magic, Witchcraft, and Science', *Philosophical Investigations*, 6 (1983), 2–36.

Crosson, F., ed. *The Autonomy of Religious Belief* (Notre Dame, Ind.: University of Notre Dame Press, 1981).

Cupitt, D. *Only Human* (London: SCM, 1985).

—— *The Sea of Faith* (London: BBC, 1984).

Darwin, C. *Autobiography*, in *Charles Darwin and T. H. Huxley: Autobiographies*, ed. G. de Beer (Oxford: Oxford University Press, 1974).

—— *The Origin of Species* (Harmondsworth: Penguin, 1968; 1st pub. 1859).

Davidson, D. 'On the Very Idea of a Conceptual Scheme', *Proceedings of the American Philosophical Association*, 47 (1973–4), 5–20.

—— 'Replies to David Lewis and W. V. Quine', *Synthèse*, 23 (1974), 345–9.

Devitt, M. *Realism and Truth* (Oxford: Blackwell, 1984).

Evans, G. R. *Augustine on Evil* (Cambridge: Cambridge University Press, 1982).

Feigl, H. and G. Maxwell, eds. *Minnesota Studies in the Philosophy of Science*, iii, (Minneapolis: University of Minnesota Press, 1962).

Ferreira, M. J. *Doubt and Religious Commitment: The Role of Will in Newman's Thought* (Oxford: Clarendon Press, 1980).

Fey, W. R. *Faith and Doubt: The Unfolding of Newman's Thought on Certainty* (Sheperdstown, W. Va.: Patmos Press, 1976).

Fine, A. 'The Natural Ontological Attitude', in *Scientific Realism*, ed. Leplin, 83–107.

Glymour, C. *Theory and Evidence* (Princeton, N J: Princeton University Press, 1980).

Griffiths, A. P., ed. *Philosophy and Practice* (Royal Institute of Philosophy Lectures, 18; Cambridge: Cambridge University Press, 1985).

Gutting, G., ed. *Paradigms and Revolutions* (Notre Dame, Ind.: University of Notre Dame Press, 1980).

Hacking, I. *Representing and Intervening* (Cambridge: Cambridge University Press, 1983).

—— ed. *Scientific Revolutions* (Oxford: Oxford University Press, 1981).

Hanson, A. and R. *Reasonable Belief* (Oxford: Oxford University Press, 1980).

Hardin, C. L. and A. Rosenberg. 'In Defense of Convergent Realism', *Philosophy of Science*, 49 (1982), 604–15.

Hare, R. M. 'Theology and Falsification', in *The Philosophy of Religion*, ed. Mitchell, 15–18.

Harman, G. 'The Inference to the Best Explanation', *Philosophical Review*, 74 (1965), 88–95.

Harré, R. *The Principles of Scientific Thinking* (London: Macmillan, 1970).

Heidelberger, M. 'Some Intertheoretic Relations Between Ptolemean and Copernican Astronomy', in *Paradigms and Revolutions*, ed. Gutting, 271–83.

Hempel, C. *Aspects of Scientific Explanation* (New York: Free Press, 1965).

—— 'Aspects of Scientific Explanation', in *Aspects of Scientific Explanation*, 331–496.

—— *Philosophy of Natural Science* (Englewood Cliffs, N J: Prentice-Hall, 1966).

—— 'Studies in the Logic of Explanation', in *Aspects of Scientific Explanation*, 245–95.

Hesse, M. *The Structure of Scientific Inference* (London: Macmillan, 1974).

Hick, J. *Evil and the God of Love*, 2nd edn. (London: Macmillan, 1977).

—— 'Remarks', in *Reason and Religion*, ed. Brown, 122–8.

Hollis, M. and S. Lukes, eds. *Rationality and Relativism* (Oxford: Blackwell, 1982).

Hudson, W. D. 'Some Remarks on Wittgenstein's Account of Religious Belief', in *Talk of God*, ed. Vesey, 36–51.

Hulme, T. E. *Speculations*, 2nd edn., ed. H. Read (London: Kegan Paul, Trench & Trubner, 1936).

Hume, D. *An Enquiry Concerning Human Understanding*, ed. L. A. Selby-Bigge, 3rd edn. (Oxford: Oxford University Press, 1975).

Kenny, A. J. P. *The God of the Philosophers* (Oxford: Clarendon Press, 1979).

Kitcher, P. *Abusing Science* (Milton Keynes: Open University Press, 1983).

Kordig, C. R. *The Justification of Scientific Change* (Dordrecht: Reidel, 1971).

Kuhn, T. S. *The Essential Tension* (Chicago: University of Chicago Press, 1977).

—— 'Logic of Discovery or Psychology of Research?', in *Criticism and the Growth of Knowledge*, ed. Lakatos and Musgrave, 1–23.

—— 'Objectivity, Value Judgment, and Theory Choice', in *The Essential Tension*, 320–39.

—— Postscript, *The Structure of Scientific Revolutions*.

—— 'Reflections on my Critics', in *Criticism and the Growth of Knowledge*, ed. Lakatos and Musgrave, 231–78.

—— 'Second Thoughts on Paradigms', in *The Essential Tension*, 293–319.

—— *The Structure of Scientific Revolutions*, 2nd edn. (Chicago: University of Chicago Press, 1970).

Lakatos, I. 'Criticism and the Methodology of Scientific Research Programmes', *Proceedings of the Aristotelian Society*, 69 (1968–9), 149–86.

Lakatos, I. (*cont.*) 'Falsification and the Methodology of Scientific Research Programmes', in *Criticism and the Growth of Knowledge*, ed. Lakatos and Musgrave, 91–196.

—— *Philosophical Papers, i. The Methodology of Scientific Research Programmes*, ed. J. Worrall and G. Currie (Cambridge: Cambridge University Press, 1978).

—— and A. Musgrave, eds. *Criticism and the Growth of Knowledge* (Cambridge: Cambridge University Press, 1970).

—— and E. Zahar. 'Why Did Copernicus's Research Programme Supersede Ptolemy's?', in *The Methodology of Scientific Research Programmes*, 168–92.

Laudan, L. 'A Confutation of Convergent Realism', *Philosophy of Science*, 48 (1981), 19–49.

—— 'The Epistemology of Light: Some Methodological Issues in the Subtle Fluids Debate', in *Science and Hypothesis*, 111–40.

—— *Progress and its Problems* (London: Routledge & Kegan Paul, 1977).

—— 'Realism without the Real', *Philosophy of Science*, 51 (1984), 156–62.

—— *Science and Hypothesis* (Dordrecht: Reidel, 1981).

—— *Science and Values* (Berkeley: University of California Press, 1984).

—— 'William Whewell on the Consilience of Inductions', in *Science and Hypothesis*, 163–91.

Leplin, J., ed. *Scientific Realism* (Berkeley: University of California Press, 1984).

Lewis, D. 'Radical Interpretation', *Synthèse*, 23 (1974), 331–44.

Lucas, J. R. *The Freedom of the Will* (Oxford: Clarendon Press, 1970).

Lyas, C. 'The Groundlessness of Religious Belief', in *Reason and Religion*, ed. Brown, 158–80.

MacIntyre, A. 'The Logical Status of Religious Belief', in *Metaphysical Beliefs*, ed. Toulmin, Hepburn and MacIntyre, 157–201.

Mackie, J. L. *Ethics* (Harmondsworth: Penguin, 1977).

—— 'Evil and Omnipotence', in *The Philosophy of Religion*, ed. Mitchell, 92–104.

—— *The Miracle of Theism* (Oxford: Clarendon Press, 1982).

—— *Problems from Locke* (Oxford: Clarendon Press, 1976).

McMullin, E. 'A Case for Scientific Realism', in *Scientific Realism*, ed. Leplin, 8–40.

—— 'Values in Science', *Proceedings of the Philosophy of Science Association*, 2 (1982), 3–28.

Masterman, M. 'The Nature of a Paradigm', in *Criticism and the Growth of Knowledge*, ed. Lakatos and Musgrave, 59–89.

Maxwell, G. 'The Ontological Status of Theoretical Entities', in *Minnesota Studies in the Philosophy of Science*, iii, ed. Feigl and Maxwell, 3–27.

Meynell, H. 'Truth, Witchcraft and Professor Winch', *Heythrop Journal*, 13 (1972), 162–72.

Mills, T. R. 'Relativism in the Analysis of Religious Belief and Language and the Construction of Theories of Meaning for Natural Languages', D.Phil. thesis (Oxford, 1982).

Mitchell, B. G. 'Faith and Reason: A False Antithesis?', *Religious Studies*, 16 (1980), 131–44.

—— *The Justification of Religious Belief* (London: Macmillan, 1973).

—— ed. *The Philosophy of Religion* (Oxford: Oxford University Press, 1971).

Musgrave, A. 'Constructive Empiricism *Versus* Scientific Realism', *Philosophical Quarterly*, 32 (1982), 262–71.

Newman, J. H. *An Essay in Aid of a Grammar of Assent*, ed. I. T. Ker (Oxford: Clarendon Press, 1985; 1st pub. 1870).

Newton-Smith, W. H. *The Rationality of Science* (London: Routledge & Kegan Paul, 1981).

—— 'Realism and Inference to the Best Explanation' (unpublished paper, 1984).

—— 'The Role of Interests in Science', in *Philosophy and Practice*, ed. Griffiths, 59–73.

O'Hear, A. *Experience, Explanation and Faith* (London: Routledge & Kegan Paul, 1984).

Phillips, D. Z. 'Belief, Change and Forms of Life: The Confusions of Externalism and Internalism', in *The Autonomy of Religious Belief*, ed. Crosson, 60–92.

—— *The Concept of Prayer* (London: Routledge & Kegan Paul, 1965).

—— *Death and Immortality* (London: Macmillan, 1970).

—— *Faith and Philosophical Enquiry* (London: Routledge & Kegan Paul, 1970).

—— *Religion without Explanation* (Oxford: Blackwell, 1976).

—— 'Religious Beliefs and Language-Games', in *The Philosophy of Religion*, ed. Mitchell, 121–42.

Pike, N. 'Hume on Evil', in *God and Evil*, ed. N. Pike (Englewood Cliffs, N J: Prentice-Hall, 1964), 85–102.

Plantinga, A. *God, Freedom, and Evil* (Grand Rapids, Mich.: Eerdmans, 1977).

—— *The Nature of Necessity* (Oxford: Clarendon Press, 1974).

—— 'Reason and Belief in God', in *Faith and Rationality* (Notre Dame, Ind.: University of Notre Dame Press, 1983), ed. A. Plantinga and N. Wolterstorff, 16–93.

Popper, K. 'Conjectural Knowledge: My Solution of the Problem of Induction', in *Objective Knowledge*, 2nd edn. (Oxford: Clarendon Press, 1979), 1–31.

Popper, K. *Conjectures and Refutations*, 4th edn. (London: Routledge & Kegan Paul, 1972).

—— *The Logic of Scientific Discovery* (London: Hutchinson, 1959).

—— 'Science: Conjectures and Refutations', in *Conjectures and Refutations*, 33–65.

—— 'Three Views Concerning Human Knowledge', in *Conjectures and Refutations*, 97–119.

Prevost, R. W. 'Theism as an Explanatory Hypothesis: Richard Swinburne on the Existence of God', D.Phil. thesis (Oxford, 1985).

Prior, A. 'The Formalities of Omniscience', *Philosophy*, 37 (1962), 114–29.

Putnam, H. 'The "Corroboration" of Theories', in *Philosophical Papers, i. Mathematics, Matter and Method* (Cambridge: Cambridge University Press, 1975), 250–69.

—— 'Explanation and Reference', in *Philosophical Papers, ii. Mind, Language and Reality*, 196–214.

—— *Meaning and the Moral Sciences* (London: Routledge & Kegan Paul, 1978).

—— 'The Meaning of "Meaning" ', in *Philosophical Papers, ii. Mind, Language and Reality*, 215–71.

—— *Philosophical Papers, ii. Mind, Language and Reality* (Cambridge, Cambridge University Press, 1975).

—— *Reason, Truth and History* (Cambridge: Cambridge University Press, 1981).

Rorty, R. *The Consequences of Pragmatism* (Brighton: Harvester, 1982).

—— 'Keeping Philosophy Pure: An Essay on Wittgenstein', in *The Consequences of Pragmatism*, 19–36.

—— *Philosophy and the Mirror of Nature* (Oxford: Blackwell, 1980).

—— 'Transcendental Arguments, Self-Reference and Pragmatism', in *Transcendental Arguments and Science*, ed. Bieri, *et al.*, 77–103.

Salmon, W. *Scientific Explanation and the Causal Structure of the World* (Princeton, N J: Princeton University Press, 1984).

Schlesinger, G. N. *Metaphysics* (Oxford: Blackwell, 1983).

Schoen, E. L. *Religious Explanations: A Model from the Sciences* (Durham, N C: Duke University Press, 1985).

Shapere, D. 'The Structure of Scientific Revolutions', *Philosophical Review*, 73 (1964), 383–94.

Sutherland, S. *God, Jesus and Belief* (Oxford: Blackwell, 1984).

Swinburne, R. *The Coherence of Theism* (Oxford: Clarendon Press, 1977).

—— *The Existence of God* (Oxford: Clarendon Press, 1979).

—— *Faith and Reason* (Oxford: Clarendon Press, 1981).

—— 'Mackie, Induction, and God', *Religious Studies*, 19 (1983), 385–91.

—— 'Simplicity and Choice of Theory in Science and Religion', unpublished paper.

Taylor, C. 'Rationality', in *Rationality and Relativism*, ed. Hollis and Lukes, 87–105.

Taylor, R. *Action and Purpose* (Englewood Cliffs, N J: Prentice-Hall, 1966).

Tennyson, Lord Alfred. *In Memoriam.*

Toulmin, S. E., R. W. Hepburn and A. MacIntyre, eds. *Metaphysical Beliefs*, 2nd edn. (London: SCM, 1970).

Turner, P. *Tennyson* (London: Routledge & Kegan Paul, 1976).

van Fraassen, B. *The Scientific Image* (Oxford: Clarendon Press, 1983).

van Inwagen, P. *An Essay on Free Will* (Oxford: Clarendon Press, 1983).

Vesey, G. N. A., ed. *Talk of God* (Royal Institute of Philosophy Lectures, 2; London: Macmillan, 1969).

Watson, G., ed. *Free Will* (Oxford: Oxford University Press, 1982).

Whewell, W. *The Philosophy of the Inductive Sciences, Founded upon Their History*, 2nd edn., 2 vols. (London: Parker, 1847).

Winch, P. 'Understanding a Primitive Society', in *Ethics and Action* (London: Routledge & Kegan Paul, 1972), 8–49.

Wittgenstein, L. *Lectures and Conversations on Aesthetics, Psychology and Religious Belief,* ed. C. Barrett (Oxford: Blackwell, 1966).

—— *On Certainty*, ed. G. E. M. Anscombe and G. H. von Wright, trans. D. Paul and G. E. M. Anscombe (Oxford: Blackwell, 1969).

—— *Remarks on Frazer's Golden Bough*, ed. R. Rhees, trans. A. C. Miles (Retford, Notts.: Brynmill Press, 1979).

—— *Tractatus Logico-Philosophicus*, trans. D. F. Pears and B. F. McGuinness (London: Routledge & Kegan Paul, 1961).

Woolf, V. *The Letters of Virginia Woolf*, iii, ed. N. Nicolson (London: Hogarth Press, 1977).

Wykstra, S. J. 'The Humean Obstacle to Evidential Arguments from Suffering: On Avoiding the Evils of "Appearance" ', *International Journal for Philosophy of Religion*, 16 (1984), 73–93.

Zahar, E. 'Why Did Einstein's Programme Supersede Lorentz's?', *British Journal for the Philosophy of Science*, 24 (1973), 95–123 and 233–62.

Index